Limitless
mind

JO BOALER

Limitless mind

Learn, lead and live without barriers

Thorsons

Thorsons
An imprint of HarperCollins*Publishers*
1 London Bridge Street
London SE1 9GF

HarperCollinsPublishers
Macken House, 39/40 Mayor Street Upper,
Dublin 1, D01 C9W8, Ireland

First published in the USA by HarperCollins*Publishers*, 2019

This edition published by Thorsons, 2019

© Jo Boaler 2019

Jo Boaler asserts the moral right to be
identified as the author of this work

Designed by Terry McGrath

A catalogue record of this book is
available from the British Library

ISBN 978-0-00-830566-6

Printed and bound in the UK using 100%
Renewable Electricity at CPI Group (UK) Ltd

This book is produced from independently certified FSC™
paper to ensure responsible forest management.
For more information visit: www.harpercollins.co.uk/green

For more information visit: www.harpercollins.co.uk/green

I dedicate these pages to the people I interviewed for the book
who opened their hearts and shared their journeys—
I could not have written this book without you.
I also dedicate this book to my two amazing daughters.
Thank you for being you, Jaime and Ariane.

CONTENTS

THE SIX KEYS

It was a sunny day and I stopped to appreciate the sunlight playing on the columns of the San Diego museum as I walked into my presentation. A flutter of nerves passed through me as I climbed the steps of the auditorium, ready to share the latest science on the ways we learn to a room packed with medical professionals. I speak regularly in front of teachers and parents but was uncertain how a different audience would relate to my latest discoveries. Would my ideas fall flat?

I needn't have worried. The response from the group of medical professionals was the same as that of the many students and educators I regularly work with. Most were surprised, some were shocked, and all could immediately see the crucial connections of these ideas to their work and lives. Several even started to see themselves in a new light. Sara—an occupational therapist—rushed up to me afterward to tell me how she dropped out of a math major many years ago when the work got difficult and she felt as if she didn't belong. She recalled an experience of being held back by damaging and incorrect beliefs about her ability. She believed, as most people do, that there were limits to what she could do.

But what if the opposite is true, and we can all learn anything? What if the possibilities to change our expertise, to develop in new directions, to form different identities as people are actually endless and continue throughout our lives? What if we wake up every day of our lives with a changed brain? This book will share evidence that our brains—and our lives—are highly adaptable, and that when people fully embrace this knowledge and change their approach to their lives and their learning, incredible outcomes result.

Almost every day I meet people who believe damaging ideas about themselves and their learning, and they come from all ages, genders, jobs, and walks of life. Typically people will tell me that they used to like math, art, English, or another subject area, but when they started to struggle, they decided they did not have the right brain for the work and gave up. When people give up on math, they also give up on all math-related subjects, such as science, medicine, and technology. Similarly, when people get the idea they cannot be a writer, they give up on all subjects in the humanities, and when people decide they are not artistic, they give up on painting, sculpture, and other aspects of the fine arts.

Every year millions of children start school excited about all they will learn, but quickly become disillusioned when they get the idea they are not as "smart" as others. Adults decide not to follow pathways they had hoped to pursue because they decide they are not good enough for them, or they are not as "smart" as other people. Thousands of employees enter meetings in the workplace anxious that they will be found out, and exposed for not "knowing enough." These limiting and damaging ideas come from inside us, but they are usually sparked by incorrect messages sent by other people, and by institutions of education. I have met so many children and adults whose lives were limited by incorrect ideas that I decided it was time to write a book dispelling the damaging myths holding people back on a daily basis; it was time to offer a different approach to life and to learning.

A large number of people are told directly that they are not a "math person" or an "English person" or an "artist" by teachers or parents. In an attempt to be helpful, adults tell young learners that a particular subject is "just not for them." This happens to some when they are children. For others, it happens later in life when they are taking college courses or interviewing for their first job. Some people are given negative messages about their potential directly; others assume it from culturally embedded ideas—that some people can achieve and some cannot.

When we learn the new science in this book and the six keys of learning I will present, our brains function differently, and we change as people. The six keys not only change people's beliefs about their reality, they change their reality. This is because as we begin to realize our potential, we unlock parts of ourselves that had been held back and start to live without limiting beliefs; we become able to meet the small and large challenges we are faced with in life and turn them into achievements. The implications of the new science are important for everyone. For teachers, leaders, and learners, the changed possibilities created by this new information are far reaching.

I am a Stanford professor of education who has spent the last few years collaborating with brain scientists, adding their knowledge of neuroscience to my knowledge of education and learning. I regularly share the new knowledge that is in this book, and invite people to think differently about problems, and it changes the way they think about themselves. I have spent the last several years focusing on mathematics, the subject with the most damaging ideas held by teachers, students, and parents. The idea that math ability (and a host of other capabilities) is fixed is a large part of the reason that math anxiety is widespread in the US and the world. Many children grow up thinking that either you can do math or you can't. When they struggle, they assume they can't. From that point on, any struggle is a further reminder of their perceived inadequacies. This affects millions of people. One study found that 48 percent of all young adults in a work-apprentice program had math anxiety;[1] other studies have found that approximately 50 percent of students taking introductory math courses in college suffer from math anxiety.[2] It is difficult to know how many people walk around in society harboring damaging ideas about their math ability, but I estimate it to be at least half of the population.

Researchers now know that when people with math anxiety encounter numbers, a fear center in the brain is activated—the *same* fear center that lights up when people see snakes or spiders.[3] As the fear center of the brain becomes activated, activity in the problem-solving centers of the brain is diminished. It is no wonder that so many people underachieve in mathematics—as soon as people become anxious about it, their brains are compromised. Anxiety in any subject area has a negative impact on the functioning of the brain. It is critical that we change the messages that are given to learners about their ability and rid education and homes of anxiety-inducing teaching practices.

We are not born with fixed abilities, and those who

achieve at the highest levels do not do so because of their genetics.[4] The myth that our brains are fixed and that we simply don't have the aptitude for certain topics is not only scientifically inaccurate; it is omnipresent and negatively impacts education and many other events in our everyday lives. When we let go of the idea that our brains are fixed, stop believing that our genetics determine our lives' pathways, and learn that our brains are incredibly adaptable, it is liberating. The knowledge that every time we learn something our brains change and reorganize comes from perhaps the most important research of this decade—research on brain plasticity, also known as neuroplasticity.[5] I will be sharing the most compelling evidence on this topic in the next chapter.

When I make the point with adults—often teachers and educators—that we should reject ideas of fixed thinking and instead see all learners as capable, these adults invariably tell me about themselves as learners. Almost all of them can recall their own experience and realize the ways in which they themselves were limited and held back. We have all been fully immersed in the damaging myth that some are smart—they have a gift or special intelligence—and some are not, and these ideas have shaped our lives.

We now know that ideas about limits to potential or intelligence are incorrect. Unfortunately, they are persistent and widespread in many cultures across the world. The good news is that when we challenge these beliefs, incredible results follow. In this book, we will upend these ingrained and dangerous self-limiting beliefs and reveal the opportunities that open up when we adopt a limitless approach. The limitless approach starts with knowledge from neuroscience and expands into a different approach to ideas and to life.

The original discovery of neuroplasticity is decades old, and the groundbreaking studies that have shown brain growth and change—among children and adults—are well established.[6] The science, however, has for the most part not made its way into classrooms, boardrooms, or homes. It has also not been translated into the much-needed ideas for learning that this book will share. Fortunately, a few pioneers who have learned about brain change have taken it upon themselves to spread the news. Anders Ericsson, a Swedish-born psychologist, is one of those people. He first became aware of the brain's incredible ability to grow and change not from the neuroscience that was emerging at the time, but from an experiment he tried with a young athlete, a runner named Steve.[7]

Ericsson set out to study the limits of people's ability to memorize a random string of digits. A study published in 1929 found that people could improve their ability to memorize. The early researchers managed to train one person to memorize thirteen random digits and another person, fifteen. Ericsson was curious to know how people improved, so he recruited Steve, whom he describes as an average Carnegie Mellon undergraduate. On the first day that Steve began working with the researchers to memorize digits, his performance was exactly average: he could remember seven numbers consistently, sometimes eight. On the following four days, Steve improved to just under nine numbers.

Then something remarkable happened. Steve and the researchers thought he had reached his limit, but he managed to push through the "ceiling" and memorize ten numbers, two more than had seemed possible. Ericsson describes this as the beginning of what became the two most surprising years of his career. Steve continued to steadily improve until he had successfully memorized and could recall a string of eighty-two random digits. Needless to say, this feat was remarkable, and it was no magic trick. This was an "average" college student unlocking his learning potential to accomplish a rare and impressive feat.

A few years later, Ericsson and his team tried the same experiment with a different participant. Renee started off much like Steve, improving her memory beyond the level of an untrained person, and she learned to memorize close to twenty digits. Then, however, she stopped improving, and after another fifty hours of training without improvement, she dropped out of the study. This set Ericsson and his team on a new quest—to work out why Steve had managed to memorize so many more digits than Renee.

This is where Ericsson began to learn more about what he called "deliberate practice." He realized that Steve's love for running had made him highly competitive and motivated. Whenever he hit what seemed like a limit, he developed new strategies to become successful. For example, he hit a barrier at twenty-four digits, so he developed a new strategy of grouping numbers into four four-digit strings. At regular intervals, Steve developed new strategies.

This approach illustrates a key takeaway—when you hit a barrier, it is advantageous to develop a new approach and come at the problem from a new perspective. Despite how logical this sounds, far too many of us fail to make adjustments in our thinking when we run into those barriers. We often decide, instead, that we cannot overcome them. Ericsson has studied human performance in many fields and concludes: "It is surprisingly rare to get clear evidence in any field that a person has reached some immutable limit on performance. Instead, I've found that people more often just give up and stop trying to improve."[8]

For the skeptics reading this—and deciding that Steve's incredible memory feat meant that he was in some way exceptional or gifted—there's more. Ericsson repeated the experiment with another runner named Dario. Dario memorized even more than Steve—more than one hundred numbers. Those who study remarkable feats performed by seemingly ordinary people find that none of the people have a genetic advantage; instead, they put in a lot of effort and practice. Not only are ideas of genetic ability misguided; they are dangerous. And yet many of our school systems are built on a model of fixed-ability thinking—limiting potential and preventing students from incredible achievement.

The six keys of learning I will share in this book create opportunities for people to excel in the learning of different subjects, but they also empower them to approach life in a different way. They allow people to access parts of themselves that were previously unavailable. Before the journey I will set out in this book, I had believed that learning about brain science and the limitless approach would change how educators approached the teaching and learning of school subjects. Through the interviews I have conducted for this book—with sixty-two people from six different countries, people of different ages, jobs, and life circumstances—I discovered the limitless approach means much more than that.

A woman who has done an enormous amount to change people's ideas about what they can do is a colleague of mine at Stanford, Carol Dweck. Dweck's research reveals that how we think about our talents and abilities has a profound impact on our potential.[9] Some people have what she has termed a "growth mindset." They believe, as they should, that they can learn anything. Others have a damaging "fixed mindset." They believe that their intelligence is more or less fixed, and although they can learn new things, they cannot change their basic intelligence. These beliefs, she has shown through decades of research, change the scope of what we can learn—and how we live our lives.

One of the important studies Dweck and her colleagues conducted took place in mathematics classes at Columbia University.[10] The researchers found stereotyping to be alive and well: young women were being given the message that they did not belong in the discipline. They also found that the message hit home only with those who had a fixed mindset. When students with a fixed mindset heard the message that math was not for women, they dropped out. Those with a growth mindset, however, protected by the belief that anyone can learn anything, were able to reject the stereotypical messages and keep going.

Throughout this book, you will learn about the importance of positive self-beliefs and ways to develop them. You will also learn about the importance of communicating positive beliefs to yourself and others, whether you are a teacher, parent, friend, or manager.

One study conducted by a group of social psychologists dramatically showed the impact of positive communication by teachers.[11] The study focused on students in high-school English classes, all of whom had written an essay. All of the students received critical, diagnostic feedback (the good kind) from their teachers, but half of the students received an extra sentence at the end of the feedback. Remarkably, the students who received the extra sentence—especially students of color—achieved at significantly higher levels in school a year later, with higher GPAs (grade point averages). So what was the sentence that those students read at the end of the feedback that caused such a dramatic result? It simply said: "I am giving you this feedback because I believe in you."

When I tell teachers about this research, I do so to show the importance of teachers' words and messages—not to suggest that they put this message at the end of every student evaluation! One teacher in a workshop raised her hand and said, "Does that mean I don't put it on a stamp?" Everyone laughed.

Studies in brain science present a very clear case for the importance of self-beliefs and the role of teachers and parents in influencing them. Yet we are living in a society where the widespread message we receive through the media on a daily basis is one of fixed intelligence and giftedness.

One of the ways children—even those as young as three—develop a damaging fixed mindset is from a small, seemingly innocuous word that is used ubiquitously. The word is "smart." Parents regularly praise their children by telling them how smart they are in order to build up their self-confidence. We now know that when we praise children for being smart, they at first think, "Oh good, I am smart," but then later when they struggle, fail, or mess up in some way, as everyone does, they think, "Oh, I am not so smart"; they end up constantly evaluating themselves against this fixed idea. It is fine to praise children, but always praise what they did and not them as people. Here are some alternatives for use in situations where you may feel the need to use the word "smart."

Fixed Praise	Growth Praise
You can divide fractions? Wow, you are smart!	You can divide fractions? That is great that you have learned how to do that.
You solved that tricky problem like that? That is so smart!	I loved your solution to the problem; it is so creative.
You have a degree in science? You are a genius!	You have a degree in science? You must have worked really hard.

I teach an undergraduate class at Stanford called "How to Learn Math" to some of the most highly achieving students in the country. They too are vulnerable to damaging beliefs. Most have been told, over many years, that they are smart, but even that "positive" message—"you are smart"—damages students. The reason it makes them vulnerable is that if they believe they are "smart" but then struggle with some difficult work, that feeling of struggle is devastating. It causes them to feel they are not smart after all and give up or drop out.

Regardless of your experience with the fixed-brain myth, the information in these pages will change your understanding of ways to raise your and other people's potential. Taking a limitless perspective is about more than a change in our thinking. It is about our being, our essence, who we are. If you live a day with this new perspective, you will know it, especially if it is a day when something bad happens, you fail at something, or you make a serious mistake. When you are limitless, you feel and appreciate such moments, but you can also move past them and even learn new and important things because of them.

George Adair lived in Atlanta after the Civil War. Originally a newspaper publisher and cotton speculator, he went on to become a highly successful real-estate developer. His success was probably spurred by an important insight that has since been widely shared: "Everything you've ever wanted is on the other side of fear." Let's think together now about ways to become limitless and move to the other side of negative beliefs and fear.

1

HOW NEUROPLASTICITY
CHANGES . . . EVERYTHING

THE SIX KEYS all have the potential to unlock different aspects of people. The first key, however, is perhaps the most critical and the most overlooked. It originates from the neuroscience of brain plasticity. Although aspects of the evidence may be familiar to certain readers, many practices in schools, colleges, and businesses are based upon ideas that are the opposite of those I will share. The result of fixed-brain thinking is that we have a nation (and world) filled with underachieving people who have been limited by ideas that could and should be changed.

LEARNING KEY #1

Every time we learn, our brains form, strengthen, or connect neural pathways. We need to replace the idea that learning ability is fixed, with the recognition that we are all on a growth journey.

Nestled in a part of California that has been described as "a piece of Tuscany transplanted into North America" is the villa that is the home of one of the world's leading neuroscientists—Michael Merzenich. It was Merzenich who stumbled upon one of the greatest scientific discoveries of our time—by accident.[1] In the 1970s, he and his team had been using the newest technologies to map out the brains of monkeys. They were making what he called "mind maps," maps of the working brain. It was exciting, cutting-edge work. The scientists hoped that the results of their studies would send ripples through the scientific community. But what Merzenich and his team discovered did not send ripples; it sent crashing waves that would go on to change people's lives profoundly.[2]

The team successfully made mind maps of the monkeys' brains and then set the maps aside to continue on with other aspects of their work. When they returned to the mind maps, they realized the monkey's brain networks, which they had sketched out in the mind maps, had changed. Merzenich himself reflected: "What we saw was absolutely astounding. I couldn't understand it."[3] Eventually the scientists drew the only possible conclusion they could—the brains of the monkeys were changing and they were changing quickly. This was the birth of what came to be known as neuroplasticity.

When Merzenich published his findings, he received pushback from other scientists. Many simply would not accept an idea they had been so certain was wrong. Some scientists had believed that brains were fixed from birth, and others that brains became fixed by the time people became adults. The evidence that adult brains were changing every day seemed inconceivable. Now, two decades later, even those who were the most vehemently opposed to the evidence from neuroplasticity research have conceded.

Unfortunately our schools, colleges, businesses, and culture have, for hundreds of years, been built around the idea that some people can and some people can't. This is why putting young students into different groups and teaching them differently made perfect sense. If individuals within a school or company weren't reaching their potential, it was not due to teaching methods or environmental factors, but to their limited brains. But now, with decades of knowledge about brain plasticity, it is time that we eradicate this damaging myth about learning and potential.

Energized by the new evidence showing brain plasticity in animals, researchers began to look at the potential of human brains to change. One of the most compelling studies of the time came from London, the city where I had my first teaching and university job. London is one of the most vibrant cities in the world—and it is always filled with millions of residents and visitors. On any day in London you will see "black cabs" zipping around the thousands of major thruways, streets, and lanes. The drivers of these iconic taxicabs hold themselves to very high professional standards. Londoners know that if they get in a black cab and tell the driver a road to find, and the driver does not know it, the driver should be reported to black-cab authorities.

Knowing all the roads in London is quite a feat—and drivers go to huge lengths to learn them. In order to become a black-cab driver, you need to study for at least four years. The most recent cab driver I traveled with told me he had studied for seven years. During this time drivers must memorize every one of the twenty-five thousand streets and twenty thousand landmarks within a six-mile radius of the centrally located Charing Cross station—and every connection between them.

This is not a task that can be accomplished through blind memorization—the drivers drive the roads, experiencing the streets, landmarks, and connections, so they can remember them. At the end of the training period, the drivers take a test that is aptly named "The Knowledge." On average, people have to take the test twelve times in order to pass it.

The extent and focus of the deep training needed by black-cab drivers caught the attention of brain scientists, who decided to study the brains of the black-cab drivers before and after the training. Their research found that, after the intense spatial training, the hippocampus of the cab drivers' brains had grown significantly.[4] This study was significant for many reasons. First, the study was conducted with adults of a range of ages, all of whom showed significant brain growth and change. Second, the area of the brain that grew—the hippocampus—is important for all forms of spatial and mathematical thinking. Researchers also found that when black-cab drivers retired from cab driving, the hippocampus shrank back down again—not from age, but from lack of use.[5] This degree of plasticity of the brain, the amount of change, shocked the scientific world. Brains were literally growing new connections and pathways as the adults studied and learned, and when the pathways were no longer needed, they faded away.

These discoveries began in the early 2000s. At around the same time, the medical world was stumbling upon its own revelations in the realm of neuroplasticity. A nine-year-old girl, Cameron Mott, was suffering from a rare condition that gave her life-threatening seizures. Doctors decided to perform a revolutionary operation, removing the entire left hemisphere of her brain. They expected Cameron to be paralyzed for many years or possibly life, as the brain controls physical movement. After the surgery, they were absolutely stunned when she started moving in unexpected ways. The only conclusion they could draw was that the right side of the brain was developing the new connections it needed to perform the functions of the left side of the brain,[6] and the growth happened at a faster rate than doctors had ever thought possible.

Since then, other children have had half of their brains removed. Christina Santhouse was eight when she had the operation—performed by neurosurgeon Ben Carson, who later would run for president. Christina went on to make the honor roll at her high school, graduate from college, and go on to achieve a master's degree. She is now a speech pathologist.

We have multiple forms of evidence, from neuroscience and from medicine, that brains are in a constant state of growth and change. Every single day when we wake up in the morning, our brains are different than they were the day before. In the next chapters you will learn ways to maximize brain growth and connectivity throughout your life.

A few years ago we invited eighty-three middle-school students to the Stanford campus for an eighteen-day math camp. They were typical students as far as their achievement levels and beliefs went. On the first day each of the eighty-three students told interviewers that he or she was "not a math person." When asked, they all named the one student in their class whom they believed to be a "math person." Unsurprisingly, it was usually the student who was quickest to answer questions.

We spent our time with the children working to change

their damaging beliefs. All students had taken a math test in their district before coming to us. We gave them the same test eighteen days later at the end of our camp. The students had improved by an average of 50 percent per student, the equivalent of 2.8 years of school. These were incredible results and further evidence of the brain's learning potential when given the right messages and forms of teaching.

When the teachers and I were working to dispel the negative beliefs the students held, we showed them images of Cameron's brain, with only one hemisphere, and told them about the operation she underwent to have half of her brain removed. We also described her recovery and how the growth of the other hemisphere had shocked doctors. Hearing about Cameron inspired our middle-school students. As they worked over the next two weeks, I often heard them say to each other, "If that girl with half a brain can do it, I know I can do it!"

So many people harbor the damaging idea that their brain is not suited to math, science, art, English, or any other particular area. When they find a subject difficult, instead of strengthening brain areas to make study possible, they decide they were not born with the right brain. Nobody, however, is born with the brain they need for a particular subject. Everyone has to develop the neural pathways they need.

Researchers now know that when we learn something, we grow the brain in three ways. The first is that a new pathway is formed. Initially the pathway is delicate and fine, but the more deeply you learn an idea, the stronger the pathway becomes. The second is that a pathway that is already there is strengthened, and the third is that a connection is formed between two previously unconnected pathways.

These three forms of brain growth occur when we learn, and the processes by which the pathways are formed and strengthened allow us to succeed in our mathematical, historical, scientific, artistic, musical, and other endeavors. We are not born with these pathways; they develop as we learn—and the more we struggle, the better the learning and brain growth, as later chapters will show. In fact, our brain structure changes with every different activity we perform, perfecting circuits so they are better suited to tasks at hand.[7]

The Fixed-Brain Message

Let's imagine how transformative this knowledge can be for the millions of children and adults who have decided they cannot learn something—and for the teachers and managers who see people struggle or fail, and decide they will never succeed. So many of us believe or were told by teachers that we were incapable of learning in a particular area. Teachers don't impart this idea to be cruel; they see their role as providing guidance on what students should or shouldn't pursue or study.

Others give this message to be comforting. "Don't worry if math isn't your thing" is, tragically, a common refrain heard by girls. Other students receive this message through faulty and outdated teaching measures, such as the separation of young children into ability groups or an emphasis on speed in learning. Whether it is through the educational system or in conversations directly with educators, far too many of us have been conditioned to believe that we don't have the capacity to learn. Once people get this terrible idea in their heads, their learning and cognitive processes change.

Jennifer Brich is the mathematics lab director at California State University San Marcos. She lectures in mathematics as well as directing the center. Jennifer works hard to dispel the damaging beliefs that her students hold about mathematics and their brains, one of very few university-level mathematics teachers doing so. Jennifer used to think that "people were born with certain talents, and you were restricted to those talents." But then she read the research on brain growth and change. Now Jennifer teaches the research on brain growth not only to her own students, but also to graduate students who teach other classes. Teaching the new science can be difficult, and Jennifer tells me that she gets a lot of pushback from people who want to believe that some people are born with math potential and others just don't have it.

A few months ago, she was sitting in her office going through emails when she heard the sound of sobbing coming from the office next door. Jennifer describes paying attention to the sound and then hearing the professor say: "It's okay. You're a female. Females have different brains than men, so you may not get this right away, and it's okay if you don't get it at all."

Jennifer was horrified and took the brave step of knocking on the door of the other professor's office. She poked her head in and asked if she could talk to the male professor.

She discussed the incorrect messages he was giving with him, which caused him to get upset and report Jennifer to the department chair. Fortunately, the department chair was a woman who also knew that his messages were incorrect and supported Jennifer.

Jennifer is taking on the myths about math and learning, and she is just the person for it. She recently told me about her own challenging experience of being discouraged by a professor when she was in grad school:

> I was a grad student, finishing my first year. I had started some research for my thesis. I was doing great; I was working really hard and getting good grades. I was in this class, it was topology, and it was really challenging for me, but I was working really hard, and I had done really well on an exam. I was really proud of myself. We had gotten the exams back, and I had gotten like a 98 or something, really close to perfect. I was so happy. Then I flipped to the back of the exam, and there was a note from my professor that said to see him after class. And I was like, "Okay, well maybe he's excited too." I was so happy and proud of myself.
>
> When I sat down in his office, we began this conversation about why I wasn't cut out for math. He wanted to know if maybe I cheated or memorized, to do so well on the exam. He pretty much told me that he didn't think that I was a mathematician, that it shouldn't be my future, and encouraged me to consider my other options.
>
> I told him I was starting my thesis that summer and what my grade point average was. So he pulled up my grades and saw that I did both undergrad and my master's there. Then he pulled up my grade record and started looking at some

of my grades. And he just kept asking me questions that all implied that I didn't earn those grades myself. It tore me apart when he did that, because he was a man I respected, someone I thought was so smart, who was very well known in the math department, very respected. A lot of the male students loved him. After that I went to my car and cried, I was so upset. I just bawled my eyes out.

My mom's a teacher, so I called my mom. When I reported the conversation, she of course got really defensive and angry. She told me to really just think about it and think about people who do well in math and why they do well. And she made me think about all these different things. I think that was the planting of the first seed that really helped me to start to understand what a growth mindset is. And following that, luckily the fierceness in me kicked in, and the feistiness, and I used that to motivate myself to do even better in that course and in my career. And I made sure to give that professor a big smile as I walked across the stage at graduation.

Jennifer's encounter tells us of a person, a professor responsible for students' lives, who believes that only some people belong in mathematics. Sadly, this professor is not alone in his incorrect thinking. The Western world, in particular, is filled with the deeply ingrained cultural belief, pervasive in all subject areas and professions, that only some people can be high-achievers. Many of us have been told this, and we have been conditioned to believe it. Once people believe that only some can reach high levels, it affects all areas of their lives and stops them from choosing fulfilling pathways. The belief that only some people can be high-achievers is insidious and damaging and prevents all of us from reaching our potential.

When teachers and others give people the idea that they don't have the brain to learn something, it is because they do not know or they refuse to accept the new scientific evidence. More often than not these are STEM (science, technology, engineering, and mathematics) teachers and professors, an issue I will return to. I think of these people as stuck inside the "fixed-brain regime." It is not surprising that so many people are stuck inside this negative place. The neuroscience showing brain growth did not become established until about twenty years ago; before that everyone believed that people were born with certain brains and those brains never changed. Many of the teachers and professors inside the fixed-brain regime have not seen the scientific evidence. University systems of reward mean that professors are most valued for publishing in scientific journals, not writing books (such as this one) for the public or sharing evidence widely. That means the most important evidence is "locked" up in journals, which are often behind paywalls, and does not get to the people who need it—in this case, educators, managers, and parents.

Changing Perceptions and Brains

It is the lack of opportunities for important knowledge to get to the people who need it that prompted Cathy Williams and me to start youcubed. This is a Stanford center and website (youcubed.org) dedicated to getting research evidence on learning to the people who need it—especially teachers and parents. We are now in a new era, and many neuroscientists and doctors are writing books and giving TED Talks in order to bring people new information. Norman Doidge is one of the people who has done a great deal to change perceptions and share the new and important brain science.

Doidge is a medical doctor who has written an incredible book with the title *The Brain That Changes Itself: Stories of Personal Triumph from the Frontiers of Brain Science*. The book is exactly what the title describes; it is filled with inspiring examples of people with severe learning disabilities or medical conditions (such as a stroke) who, although written off by educators and doctors, have undergone brain training and recovered completely. Doidge works to shatter a number of myths in the book, such as the idea that separate brain areas are compartmentalized and don't communicate or work together and, most important, the idea that brains don't change. Doidge describes the "dark ages" when people believed that brains were fixed, says that he is unsurprised that people are slow to understand the plasticity of the brain, and suggests that it will take an intellectual "revolution" for them to do so.[8] I agree, because over the last few years in my teaching about the new brain science I have met many people who seem unwilling to make the shift in their understanding of the brain and human potential.

The vast majority of schools are still inside the fixed-brain regime. Schooling practices have been set over many years and are very difficult to change. One of the most popular is tracking, a system in which students are placed in groups based on their supposed ability and then taught together in those groups. A study in Britain showed that 88 percent of students placed into tracks at the age of four remained in the same track for the rest of their school lives.[9] This horrific result does not surprise me. Once we tell young students they are in a lower-track group, their achievement becomes a self-fulfilling prophecy.

The same is true when teachers are told which tracks students are in; they treat students differently whether they intend to or not. Similar results were found in a study of nearly twelve thousand students from kindergarten to third grade in more than twenty-one hundred schools in the US.[10] None of the students who started out in the lowest reading group ever caught up to their peers in the highest group. Such policies of placing students in groups based on their supposed level of ability may be defensible if it resulted in higher achievement for the low-, middle-, or high-achieving students, but it does not.

Studies of schools' tracking policies in reading show that those schools that use tracked reading groups almost always score lower on average than schools that do not.[11] These results are echoed in mathematics. I have compared students learning mathematics in middle and high schools in England and the US, and in both school levels and countries the schools that taught students in mixed achievement groups outperformed those that used ability groups.[12]

San Francisco Unified is a large and diverse urban school district whose school board voted, unanimously, to remove advanced classes until eleventh grade. This prompted a lot of controversy and opposition from parents, but within two years, during which all students took the same mathematics classes until tenth grade, algebra failure rates fell from 40 to 8 percent of students in the district and the number of students taking advanced classes after tenth grade went up by one-third.[13]

It is hard to imagine that the teaching practices of the district teachers changed dramatically in two years, but what did change were the opportunities students received to learn and the ideas students believed about themselves. All students, instead of some students, were taught high-level content—and the students responded with high achievement. International studies of achievement in different countries across the world show that countries that use tracking the latest and the least are the most successful. The US and the UK, two countries in which I have lived and worked, have two of the most highly tracked systems in the world.

Nobody knows what children are capable of learning, and the schooling practices that place limits on students' learning need to be radically rethought. Someone whose story illustrates most clearly for me the need to change our expectations of young children is Nicholas Letchford. Nicholas grew up in Australia, and in his first year of school his parents were told that he was "learning disabled" and had a "very low IQ." In one of his mother's first meetings with teachers, they reported that he was the worst child they had seen in twenty years of teaching. Nicholas found it difficult to focus, make connections, read, or write. But over the next few years Nicholas's mother, Lois, refused to believe that her son could not learn, and she worked with Nicholas, teaching him how to focus, connect, read, and write. The year 2018 was an important one for Lois Letchford. It was the year that she published a book describing her work with Nicholas, called *Reversed*,[14] and it was also the year Nicholas graduated from Oxford University with a doctoral degree in applied mathematics.

Research and science have moved beyond the fixed-brain

era, but fixed-brain schooling models and limited-learning beliefs persist. As long as schools, universities, and parents continue to give fixed-brain messages, students of all ages will continue to give up on learning in areas that could have brought them great joy and accomplishment.

The new brain science showing that we have unlimited potential is transformative for many—and that includes those diagnosed with learning disabilities. These are individuals who are born with or develop, through injury or accident, physical brain differences that make learning more difficult. For many years, schools have traditionally put such students into lower-level classes and worked around their weaknesses.

Barbara Arrowsmith-Young takes an entirely different approach. I was fortunate to meet Barbara on a recent visit to Toronto, during which I toured one of the incredible "Arrowsmith" schools she has set up. It is impossible to spend time with Barbara and not realize that she is a force to be reckoned with; she is passionate not only about sharing her knowledge of the brain and how we develop it, but in using her knowledge to change the neural pathways of those diagnosed with special educational needs through targeted brain training.

Barbara is someone who was herself diagnosed with severe learning disabilities. As she was growing up in Toronto in the 1950s and 1960s, she and her family knew she was brilliant in some areas, but they were told she was "retarded" in others. She had trouble pronouncing words and could not engage in spatial reasoning. She could not follow cause-and-effect statements, and she reversed letters. She was able to understand the words "mother" and "daughter," but not the expression "mother's daughter."[15] Fortunately for Barbara, she had an amazing memory and was able to memorize her way through school and hide what she knew was wrong.

As an adult her own disabilities prompted her to study child development, and eventually she came across the work of Alexander Luria, a Russian neuropsychologist who had written about stroke victims who had trouble with grammar, logic, and reading clocks. Luria worked with many people with brain injuries, produced an in-depth analysis of the functioning of various brain regions, and developed an extensive battery of neuropsychological tests. When Barbara read Luria's work, she realized she herself had brain injuries, became quite depressed, and started to consider suicide. But then she came across the first work on neuroplasticity and realized that particular activities could produce brain growth. She began months of detailed work on the areas she knew she was weakest in. She made herself hundreds of cards with clock faces and practiced so much she was reading them faster than "regular" people. She started to see improvements in her symbolic understanding and for the first time began grasping grammar, math, and logic.

Now Barbara runs schools and programs that give brain training to students diagnosed with learning differences. Chatting with Barbara on my visit, I found it hard to imagine that this woman herself had had such severe disabilities in the past, as she is an impressive communicator and thinker. Barbara has developed over forty hours of tests that diagnose students' brain strengths and weaknesses and a range of targeted cognitive exercises that enable students to develop brain pathways. Students come to her Arrowsmith schools with severe disabilities and leave without them.

When I visited one of the Arrowsmith schools for the first time, I saw students sitting at computer screens intensely concentrating on their cognitive tasks. I asked Barbara if the students were happy doing this, and she replied that students stay motivated because they can feel the effects of the program very quickly. Many of the students I spoke to talked in the same terms—after they started on the cognitive tasks, they felt a "fog lifting" and were able to make sense of the world. When I visited the Arrowsmith school for the second time, I sat and talked to some adults going through the program.

Shannon was a young lawyer who had become concerned after criticism for the length of time it took her to produce her work, as people typically pay for lawyers' time by the hour. She was referred to Arrowsmith and decided to enroll for a summer. When I met her, a few weeks into the program, she told me that it was already "life changing" for her. Not only was Shannon thinking a lot more efficiently, but she was able to make connections she had not been able to make before. She was even making sense of events that had happened in her past, even though she had not been able to make sense of them at the time. Shannon, like the others, talked about a "fog lifting" from her mind; she said she used to be a passenger in conversations, but now "everything is clear" and she is able to participate fully.

Barbara not only offers brain training for students who go to Toronto and enroll in the school; she has now developed a program that educators can be trained in and take back to their schools. Some students stay in the program for a few months, some for a few years, and now a remote program is being developed for students to work in different locations.

Barbara is somebody who is leading the world in her brain-training approach. Like many groundbreakers, she has had to endure critiques from the people who do not accept the idea of neuroplasticity or that brains can be exercised and developed, but she has continued fighting for the rights of students who have been made to believe they are "broken."

Most of the students who contact Arrowsmith have been given the idea there is something terribly wrong with them, and many of them have been rejected by the school system. They leave Arrowsmith transformed. One of the results of my visits to the school was that I became determined to help spread the news of what is possible with brain training and share the Arrowsmith methods with our army of teachers and parents who follow youcubed (they call themselves youcubians). As mentioned, the approach of special education in schools has been to identify students' weaknesses and teach around them, essentially teaching to their strengths. Arrowsmith's approach is the opposite. The teachers work to identify brain weaknesses and then teach to them—building up the brain pathways and connections that students need. My hope is that all students with learning differences will be exposed to brain training and freed from the labels and limits they have been forced to live with, replacing these instead with hope engendered by a transformed brain.

Many amazing individuals who were written off and told not to pursue particular studies have excelled in them. Dylan Lynn was diagnosed as having dyscalculia, a particular brain condition that makes learning mathematics hard. But Dylan refused to accept that she could not learn math and pursued and achieved a degree in statistics. She did this by refusing to listen to all the people who told her to drop her mathematics courses, instead working out her own approach to mathematics. Dylan now collaborates with Katherine Lewis, a professor at the University of Washington, in telling her story to inspire other learners who were told they could not achieve their desired goal.[16]

It is time to recognize that we cannot label children and have low expectations for them. This is true regardless of any diagnosed learning difference. As we ourselves are learning in these pages, the most notable quality of our brains is their adaptability and potential for changing and growing.

In addition to children with genuine learning disabilities, many other students are either told or made to believe they have a learning disability when they do not—particularly when it comes to mathematics. For decades, teachers everywhere have identified children who do not memorize math facts as well as their classmates and labeled them as having a deficiency or disability.

One study, conducted by neuroscientist Teresa Iuculano and her colleagues at Stanford School of Medicine, clearly shows the potential of children's brains to grow and change as well as the danger of misdiagnosing students.[17] The researchers brought in children from two groups—one group had been diagnosed as having mathematical learning disabilities and the other consisted of regular performers. The researchers used MRI scans to look at the brains of the children when they were working on math. They found actual brain differences. This is where it gets interesting. The difference was that the students identified as having disabilities had more brain regions lighting up when they worked on a math problem.

This result is counterintuitive, for many people think

that students with "special needs" have less going on in their brains, not more. However, we do not want all of the brain lighting up when we work on mathematics; we want a few focused areas to light up. The researchers dug further and gave one-on-one tutoring to both sets of students—those who were regular performers and those identified as having a mathematical learning disability. At the end of the eight weeks of tutoring, not only did both sets of students have the same achievement; they also had the exact same brain areas lighting up.

This is one of many important studies showing that after a short period of time—research interventions are often eight weeks long—brains can be completely changed and rewired. The "learning disabled" students in this study developed their brains to an extent that allowed them to function in the same way as "regular performers." Let's hope they returned to school and lost their "mathematical disability" labels. Just imagine how everything could change for those young children in school and in life.

High-Achieving Students

The importance of knowing about brain growth is not limited to students diagnosed with learning differences. It extends across the entire achievement spectrum. Students come to Stanford with a history of school success; often they have only ever received As in school. But when they struggle in their first math (or any other) class, many decide they cannot learn the subject and give up.

As mentioned, for the last several years I have been working to dispel these ideas with students by teaching a class called "How to Learn Math." The class integrates the positive neuroscience of learning with a new way of seeing and experiencing math. My experience of teaching this class has been eye-opening. I have met so many undergrads who are extremely vulnerable and too readily come to believe they don't belong in STEM subjects. Unfortunately, they are almost always women and people of color. It is not hard to understand why these groups are more vulnerable than white males. The stereotypes that pervade our society based on gender and color run deep and communicate that women and people of color are not suited to STEM subjects.

One study published in the premier journal *Science* showed this powerfully.[18] Sarah-Jane Leslie, Andrei Cimpian and colleagues interviewed university professors in different subject areas to see how prevalent the idea of a "gift" was—the concept that you need a special ability to be successful in a particular field. Their results were staggering. They found that the more prevalent the idea of a gift was in any academic field, the fewer women and people of color were in that field. This held across all thirty subjects they looked at. The following graphs show the relationships the researchers uncovered; the top chart (A) shows the science and technology subjects, and the chart below (B) shows the arts and humanities subjects.

The question I always ask when I see data like this is: If the idea of giftedness is harmful to adults to this extent, what does it do to young children?

The idea of giftedness is not only inaccurate and damaging; it is gender and racially biased. We have many different forms of evidence showing that those who believe in fixed brains and giftedness also believe that boys, men, and cer-

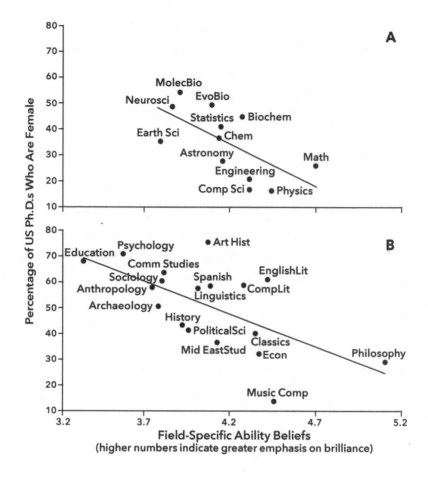

tain racial groups are gifted and girls, women, and other racial groups are not.

One of the forms of evidence that shows this clearly was collected by Seth Stephens-Davidowitz, who focused his attention on google searches.[19] His study revealed something very interesting and disturbing. He found that the most commonly googled word following "Is my two-year-old son . . ." is "gifted." He also found that parents search the words "Is my son gifted?" two and a half times more than the words "Is my daughter gifted?" This is despite the fact that young children of different genders have equal potential.

Sadly, the problem is not limited to parents. Daniel Storage and his colleagues conducted analyses of anonymous reviews on RateMyProfessors.com, and they found that students were twice as likely to call male rather than female professors "brilliant" and three times as likely to call male rather than female professors "geniuses."[20] These and other studies show that ideas of giftedness and genius are intertwined with racist and sexist assumptions.

I am convinced that the majority of people who have gender or racial biases do not think about them consciously or perhaps even realize they have them. I also contend that if we were to dispel the idea that some people are "naturally" gifted and instead recognize that everyone is on a growth journey and can achieve amazing things, some of the most insidious biases against women and people of color would disappear. This is needed in the STEM fields more than anywhere else; it is no coincidence that STEM subjects evidence the strongest fixed thinking and the starkest inequities in participation.

Part of the reason so many students are dissuaded from thinking they are capable of learning math is the attitudes of the teachers and professors who teach them. I have now met a few amazing mathematicians who devote large parts of their lives to dispelling the elitist ideas that pervade mathematics. University mathematician Piper Harron, one of my own heroes, is one of those people. On her website, called The Liberated Mathematician, she writes: "My view of mathematics is that it is an absolute mess which actively pushes out the sort of people who might make it better. I have no patience for genius pretenders. I want to empower the people."[21] It is wonderful to have voices like Piper's to help dispel the myths about who can achieve in mathematics.

Unfortunately, there remain too many academics and teachers who continue to transmit false elitist ideas and willfully and openly state that only some people can learn their subjects. Just last week I learned of two examples that are typical. A community college professor started her class by telling the students that only three of them would make it, and a high-school math teacher in my local school district announced to his eager fifteen-year-old students who were placed in his high-level math class: "You may think you are hot shit, but no one gets above a C in this class." These are the words of elitists who revel in the low number of students who are successful in their classes, as they think it shows that they are teaching really difficult content. It is this sort of thinking and speaking to students that has kept so many amazing people from pursuing pathways that would have been rewarding for them. Such ideas harm people, and they harm the disciplines, because access is denied to the diverse thinkers who would have provided beneficial insights and breakthroughs in these fields.

One of these thinkers was the incredible mathematician Maryam Mirzakhani. The story of Maryam's life and work appeared in newspapers worldwide when she became the first woman in the world to win the coveted Fields Medal—the equivalent of the "Nobel Prize" for mathematicians. Maryam grew up in Iran and, like many others, was not inspired by school math classes. In seventh grade, Maryam was told by her math teacher that she was not good at math. Fortunately for the world, Maryam had other teachers who believed in her.

At age fifteen things changed for Maryam when she signed up for a problem-solving class at Sharif University in Tehran. She loved mathematical problem solving and went on to study advanced mathematics. During her PhD studies she proved several previously unproved theories in mathematics. Her approach was different from that of many mathematicians, and her work almost entirely visual. The field would be narrower—less rich, visual, and connected—without Maryam's contribution, one that could so easily have been lost if she had listened to the teacher who told her that she was bad at math.

When Maryam came to Stanford, we found many occasions to meet and discuss mathematics learning, and I enjoyed chairing a PhD exam for one of her students. At age forty, she tragically died. The world lost an incredible woman, although her ideas will always live on and continue to broaden mathematics.

The American Mathematical Society recently devoted the November issue of its journal to Maryam, and one of those reflecting on Maryam's amazing contribution to mathematics was Jenya Sapir, the doctoral student whose thesis defense I chaired, now a mathematician herself. Here are her reflections on Maryam:

> Maryam would paint beautiful, detailed landscapes in her
> lectures. If she were giving a talk about concepts A, B, and C,
> she would not just explain that A implies B implies C.
> Rather, she would paint a mathematical landscape where
> A, B, and C lived together and interacted with one another
> in various complicated ways. More than that, she made it
> seem like the rules of the universe were working harmoni-
> ously together to make A, B, and C come about. I was often
> amazed by what I imagined her inner world to be like. In my
> imagination it contained difficult concepts from disparate
> fields of mathematics all living together and influencing one

another. Watching them interact, Maryam would learn the
essential truths of her mathematical universe.[22]

The world is filled with cases of people who think
differently—often more creatively—and are dissuaded from
pursuing careers in sports, music, academics, and many
other fields. Those who persist despite the negative messages
they receive often go on to achieve incredible feats.

But how many are there who do not go forward, who believe negative judgments and who turn away from fields and
dreams? One of the people who thought differently and received extensive rejection is the author of the *Harry Potter* series, J. K. Rowling, now the most successful author in history.
Shortly after the death of her mother, she was at a very low
point in her life; she was recently divorced, a single mother,
living in poverty, but she focused on something she cared
deeply about—writing. Rowling (also called Jo) sent her *Harry
Potter* manuscript to twelve different publishers, all of whom
rejected it.

She began to lose confidence in her book when the editor
at Bloomsbury Publishing sat down to read the book; she
also gave it to her eight-year-old daughter. The young reader
loved it and encouraged her mother to publish it. Rowling's
books have now sold over 500 million copies, and she is a role
model for any who face rejection yet believe in their ideas.
Today she actively works to end poverty and support children's welfare. I love many of her words, but this is perhaps
my favorite quote of all:

> *It is impossible to live without failing at something, unless*
> *you live so cautiously that you might as well not have lived at*
> *all—in which case, you fail by default.*

The Problems of Giftedness

The teachers, professors, and parents who maintain that only some people can learn subjects are all reflecting the misinformation of the fixed-brain era. It is perhaps not surprising that so many people still cling to the idea of fixed brains, as most of them lived during the years when this was all anyone knew. The fixed-brain myths have been devastating for students of all ages who have been written off in schools, classrooms, and homes, millions of children who have been made to believe they cannot achieve. But there is another side to this story as well. Fixed-brain thinking has also had negative consequences for the students who have been held up as being "gifted." This may seem nonsensical—how can being labeled as gifted possibly harm anyone? I have already mentioned the research showing that the idea of giftedness— that you need some inherited gene to do well—is harmful for women and students of color, but how does it harm individuals who are given the label?

A few months ago, I was contacted by a filmmaker who was making a film on giftedness with a social justice angle. That, I thought, sounded interesting, so I looked at the trailer he sent me. I was disappointed to find that his argument was that more students of color should be identified as gifted. I understand the motives for such a film, as there are serious racial disparities in gifted programs. But there was a larger issue at play, and that was the continued practice of fixed-brain labeling.

I decided in those moments to make my own film, with the help of my youcubed team and an amazing filmmaker, Sophie Constantinou, from Citizen Film. I asked the Stanford students I knew to reflect upon their experiences of being labeled as "gifted."[23] The twelve Stanford students who speak in the film give a consistent message—they received advantages, but at some costs. The students talk about feeling that they had a fixed thing inside them, and when they struggled, they thought it had "run out." They say they learned that they could not ask questions; they could only answer other people's questions. They say that they tried to hide any struggles, in case people found out that they did not have a "gift." At the end a student named Julia strikingly says, "If I grew up in a world where no one was labeled as gifted, I would have asked a lot more questions."

The gifted movement has the worthy ideal of ensuring that high-achieving students get a rich and challenging environment, which I agree is needed. But they have done so by perpetuating an idea that some students are worthy of this because they have a fixed "gift"—like a present they have been given. Although the programs point out that some students need especially challenging material because they have reached an elevated point, they omit the fact that others can also reach that point if they work hard. The message is that some people are born with something that others cannot achieve. This, in my view, is damaging, both for those who get the idea they have no gift and for those who get the idea they have a fixed brain.

One of the reasons that it can be damaging to receive the gifted label is that you do not expect to struggle, and when you do, it is absolutely devastating. I was reminded of this when chatting with my education students at Stanford last summer. I was explaining the research on brain growth and the damage of fixed labels when Susannah raised her hand and sadly said, "You are describing my life."

Susannah went on to recall her childhood, when she was a top student in math classes. She had attended a gifted program and had been told frequently that she had a "math brain" and a special talent. She went on to enroll as a math major at UCLA, but in the second year of the program she took a class that was challenging and that caused her to struggle. At that time, she decided she did not have a math brain after all, and she dropped out of the program. What Susannah did not know is that struggle is the very best process for brain growth (more on that later) and that she could grow the neural pathways she needed to learn more mathematics. If she had known that, Susannah would probably have persisted and graduated with a math major. This is the damage that is caused by fixed-ability thinking.

The story Susannah told me relayed her experience of being labeled as gifted, with a "math brain," and the ways this fixed labeling led her to drop the subject she loved. This could be repeated with any subject—English, science, history, drama, geography—anything. When you are valued for having a brain that you did not develop, one you were just given at birth, you become averse to any form of struggle and start to believe you do not belong in areas where you encounter it. Because of my field of specialty, I have met many people who have dropped out of STEM subjects because they thought they did not have the right brain, but the problem is not limited to STEM subjects. It comes about whenever people are led to believe that their intellect is fixed.

Although I decry the labels given to students—of giftedness or the opposite—I do not maintain that everyone is born the same. At birth everyone has a unique brain, and there are differences between people's brains. But the differences people are born with are eclipsed by the many ways people *can* change their brains. The proportion of people born with brains so exceptional that those brains influence what they go on to do is tiny—less than 0.001 percent of the population. Some have brain differences that are often debilitating in some ways, such as those on the autism spectrum, but productive in other ways. Although we are not born with identical brains, there is no such thing as a "math brain," "writing brain," "artistic brain," or "musical brain." We all have to develop the brain pathways needed for success, and we all have the potential to learn and achieve at the highest levels.

Bestselling author Daniel Coyle, who has spent a lot of time in "talent hotbeds," agrees. He has interviewed teachers of the most "talented"—the people Coyle describes as having worked in particularly effective ways. Their teachers say that they see someone they regard as a "genius" at a rate of one person per decade.[24] To decide that 6 percent of students in every school district have a brain difference that means they should be siphoned off and given special treatment is ludicrous. Anders Ericsson has studied IQ and hard work for decades and concludes that the people regarded as geniuses—people like Einstein, Mozart, and Newton—"are made, not born," and their success comes from extraordinary hard work.[25] Importantly, we should communicate to all students that they are on a growth journey, and there is nothing fixed about them, whether it is called a "gift" or a disability.

We are no longer in the fixed-brain era; we are in the brain-growth era. Brain-growth journeys should be celebrated, and we need to replace the outdated ideas and programs that falsely deem certain people more capable than others, especially when those outdated labels become the source of gender and racial inequalities. Everybody is on a growth journey. There is no need to burden children or adults with damaging dichotomous thinking that divides people into those who can and those who cannot.

The idea that women have to work hard to be successful whereas men are naturally brilliant was a notion I myself encountered in high school—not from my math teacher, but from my physics teacher. I remember it clearly. It was at the time when all students took a practice exam, known as a "mock exam," in preparation for the high-stakes exam all students take at age sixteen in England. Eight students—four girls and four boys—received borderline scores, and I was one of them. At this point my physics teacher decided that all the boys had achieved their scores without trying, but all the girls had achieved their scores from working hard—and so they could never do any better. As a result, he put all the boys in for the higher exam and the girls were entered for the lower exam.

Since I did hardly any work in high school (I was bored a lot of the time from just having to memorize facts) and skated by with minimal effort, I knew he was wrong about the girls having worked harder. I told my mother about the teacher's decision based on gender. My mum, being the feminist she was, complained to the school, so they grudgingly put me in for the higher exam, telling me it was a stupid risk on my part, because the only grades given for the higher exam were A, B, C, or failing. I said I would take the risk.

Later that summer I received my result—an A. I was fortunate that I had a parent who got the sexist decision the teacher made overturned, and countering his thinking gave me a reason to work especially hard for the exam. The unfortunate impact for me, however, was that I decided I would not go any further in physics. I just did not want any more to do with the man (who was the department chair) or the subject.

Luckily, I did not receive such sexist dissuasion in math, and some of my best and highest-level math teachers and professors were women. I chose to take advanced mathematics instead—I took all the sciences at advanced levels except physics. This is an example of the particularly insidious impact that men like my physics teacher have when they limit pathways based on gender (or race or other characteristics).

A group of young women recently shared with me their experience of approaching their mathematics professor with a question after one of their first classes at a top university. When they asked their question, the professor said it was too basic and they should take a class at the local community college. The women, all African American students, decided at that moment to leave STEM subjects for good. They had experienced enough of these messages and, like many other students before them, they walked away.

Mathematics is, of course, not the only subject that fuels damaging ideas about who can achieve. Art, English, music, sports—all of these are areas where students are initially interested until they begin to struggle and decide they don't have the right kind of brain (or body). In all cases when students get these damaging ideas, some portion of their future potential is foreclosed. And not only in school. Fixed ideas about potential impact their work lives as well.

I have now talked with many professionals who tell me

that before they learned about brain science, they were too nervous to offer ideas in meetings, in case they were wrong, and they were always living in fear of being judged. This is not surprising, as we have grown up in a fixed-brain world that judges everyone on their "smartness." Many of us have grown up feeling judged for everything, often feeling "not good enough" and worrying about being found out. When people let go of fixed-brain ideas, they become unlocked, especially when they combine this knowledge with other findings from neuroscience that we will explore shortly.

Workers suffer the effects of fixed-brain thinking, but often managers do as well. Managers in companies are just as likely to write off an employee as not having a good brain or being smart enough. If, instead, managers saw the limitless potential of the people they work with, they would talk to them differently and open up opportunities rather than close them down. Instead of deciding that some workers are of limited value, managers might decide that they could be given different opportunities for learning—some may need something to read or study or build (more on this in later chapters). This would change the ways many companies operate and empower many more employers to create important ideas and products.

The first step in living a limitless, unlocked life is to know brains are constantly reorganizing, growing, and changing. Remembering that every day of our lives, we wake up with a changed brain. In every moment of our lives our brains have opportunities to make connections, to strengthen pathways, and to form new pathways. When we face a challenging situation, rather than turn away because of fear of not being good enough, we should dive in, knowing that the situation presents opportunities for brain growth. As we start to recognize the huge implications of the adaptability of our brains, we will start to open our minds, and live differently. The key information that will enable our new pathways and approaches to be enhanced further will be shared in the remaining chapters.

2

WHY WE SHOULD LOVE MISTAKES, STRUGGLE, AND EVEN FAILURE

O UR LIVES are filled with mistakes. We make them all the time, and they are simply part of everyday life. Even though mistakes sometimes make no difference or end up producing fortuitous results, most of us instinctively respond to mistakes by mentally beating ourselves up and feeling terrible. It is not surprising that large sections of the population respond negatively to mistakes. Most of us have grown up with the idea that mistakes are bad, especially if we attended test-driven schools, where we were frequently marked down for making mistakes, or our parents punished mistakes with harsh words and actions. This is unfortunate, and this is why.

LEARNING KEY #2

The times when we are struggling and making
mistakes are the best times for brain growth.

When we are willing to face obstacles and make mistakes in the learning process, we enhance neural connections that expedite and improve the learning experience. The research on the positive impact of mistakes and struggle is emerging from both neuroscience[1] and behavioral studies of high-achieving people.[2] Some of this work is counterintuitive, as we have believed for so long in the absolute necessity that everything be "correct." Releasing people from the idea that they must always be correct and not make any mistakes turns out to be transformative.

The Science of Mistakes

I first became aware of the positive impact of mistakes when I was hosting a workshop for teachers and Carol Dweck, the pioneer of mindset research, joined us. The teachers attending the workshop that day had gathered in a large group and listened attentively to Carol. She announced that every time we make mistakes, synapses fire in the brain, indicating brain growth. All the teachers in the room were shocked, as they had all been working under the premise that mistakes are to be avoided. Carol was drawing from work that has researched the brain's response when we make mistakes, particularly investigating the different ways brains respond when people have a growth or a fixed mindset.[3]

Jason Moser and his colleagues extended Carol's work investigating the brain's response when we make mistakes. Moser and his team found something stunning. They had asked participants to take tests while they monitored the participants' brains with MRI technology. They looked at the scans when people got questions correct and when they got them incorrect. The researchers found that when people made mistakes, brains were more active, producing strengthening and growth, than when people got work correct.[4] Neuroscientists now agree that mistakes positively contribute to the strengthening of neural pathways.

This learning key is particularly significant because most teachers design classes so that everyone is successful. Curricula and textbooks are designed with trivial, unchallenging questions, so that students will get a high percentage of answers correct. The common belief is that getting most answers correct will motivate students toward greater success. Here's the problem, though. Getting questions right is not a good brain exercise.

For students to experience growth, they need to be working on questions that challenge them, questions that are at the edge of their understanding. And they need to be working on them in an environment that encourages mistakes and makes students aware of the benefits of mistakes. This point is critical. Not only should the work be challenging to foster mistakes; the environment must also be encouraging, so that the students do not experience challenge or struggle as a deterrent. Both components need to work together to create an ideal learning experience.

Author Daniel Coyle studied "talent hotbeds," places producing a larger proportion than normal of high-achievers, and concludes that achievement comes not from any natural-born ability, but rather from a special kind of work and practice. He has studied examples of those who excel at learning in music, sports, and academic subjects. His research reveals that all of the people who achieved at very high levels engaged in a particular kind of practice that caused the coating of brain pathways with myelin.

Our brains function through an interconnected web of nerve fibers (including neurons), and myelin is a form of insulation that wraps around fibers and increases their signal strength, speed, and accuracy. When we revisit an idea or kick a soccer ball, myelin coats the neural pathways involved, optimizing the particular circuits and making our movements and thoughts more fluid and efficient in the future. Myelin is vital to the learning process. Most learning takes time, and myelin aids the process by reinforcing signals and slowly strengthening pathways. Coyle gives a number of examples of the highest-achieving mathematicians, golfers, soccer players, and pianists practicing their craft and describes the role of myelin in wrapping layers of insulation around their neural circuits. He characterizes the world's experts as having "super-duper pathways" wrapped in layer upon layer of myelin, which makes them very effective.

So how do we all develop "super-duper pathways"? This occurs when people are working at the edge of their understanding, making mistake after mistake in difficult circumstances, correcting mistakes, moving on and making more mistakes—constantly pushing themselves with difficult material.

Coyle starts his book with an interesting story of learning. He describes the case of a thirteen-year-old girl he calls Clarissa, who is learning the clarinet. Clarissa, he says, has no musical "gifts," lacks a "good ear," and has only an average sense of rhythm and subpar motivation—yet she became famous in music circles, because she managed to accelerate her learning by ten times, according to the calculations of music psychologists. This amazing learning feat was captured on video and has been studied by music experts. Coyle describes watching the video of Clarissa practicing and suggests that the video be given a title of "The Girl Who Did a Month's Worth of Practice in Six Minutes." He describes the practice session this way:

> *Clarissa draws a breath and plays two notes. Then she stops. She pulls the clarinet from her lips and stares at the paper. Her eyes narrow. She plays seven notes, the song's opening phrase. She misses the last note and immediately stops, fairly jerking the clarinet from her lips. . . . She starts over and plays the riff from the beginning, making it a few notes farther into the song this time, missing the last note, backtracking, patching in the fix. The opening is beginning to snap together—the notes have verve and feeling. When she's finished with this phrase, she stops again for six long seconds, seeming to replay it in her mind, fingering the clarinet as she thinks. She leans forward, takes a breath, and starts again.*
>
> *It sounds pretty bad. It's not music; it's a broken-up, fitful, slow-motion batch of notes riddled with stops and misses. Common sense would lead us to believe that Clarissa is failing. But in this case common sense would be dead wrong.*[5]

A music expert watching the video commented on Clarissa's practice, saying it was "amazing" and, "If somebody could bottle this, it'd be worth millions." Coyle points out: "This is not ordinary practice. This is something else: a highly targeted, error-focused process. Something is growing, being built. The song begins to emerge, and with it, a new quality within Clarissa."[6]

In each of the learning cases Coyle reviews, he says that the learner has "tapped into a neurological mechanism in which certain patterns of targeted practice build skill. Without realizing it, they have entered a zone of accelerated learning that, while it can't quite be bottled, can be accessed by those who know how. In short, they've cracked the talent code."[7]

One of the significant characteristics of the highly effective learning described is the presence of mistakes and the role of struggle and error in transforming people from beginners into experts. This is consistent with the brain research showing increased brain activity when people struggle and make mistakes and decreased activity when they get work correct.[8] Unfortunately, most learners think they should always be getting work correct, and many feel that if they make mistakes or struggle, they are not good learners—when this is the very best thing they can be doing.

Practice is important for the development of any knowledge or skill. Anders Ericsson helped the world understand the nature of expert performance and found that most world-class experts—pianists, chess players, novelists, athletes—practiced for around ten thousand hours over twenty years. He also found that their success was not related to tests of intelligence but to the amount of "deliberate practice" they undertook.[9] Importantly, although people succeed because they are trying hard, the people who become experts are trying hard in the right way. A range of different researchers describe effective practice in the same way—people pushing at the edge of their understanding, making mistakes, correcting them, and making more.

A Different View of Struggle

Every four years an international test of mathematics and science called TIMSS (Trends in International Mathematics and

Science Study) is conducted in fifty-seven countries. In the last round of testing, Singapore was the highest-performing country in mathematics. The information from such tests is not very useful if we do not know what approach countries use to bring about their results. Accordingly, a group of researchers studied the nature of math teaching by going into classrooms and recording a representative sample of the teaching in seven countries. This teaching study uncovered a number of noteworthy outcomes.[10] One finding was that the mathematics curriculum in the US is "a mile wide and an inch deep" compared to the curriculum in more successful countries.

Japan has always scored well in mathematics—it has always finished in one of the top-five TIMSS positions—and was one of the countries visited in the study. The researchers found that Japanese students spent 44 percent of their time "inventing, thinking, and struggling with underlying concepts," whereas students in the US engaged in this kind of behavior less than 1 percent of the time.

Jim Stigler, one of the authors of the study, writes that the Japanese teachers *want* the students to struggle—and recalls the times when they would purposely give the wrong answer so that students would go back and work with foundational concepts. In my thousands of observations of classrooms over many years in the US and the UK, I have never seen this kind of practice; more typically I have seen teachers who seem to want to save students from struggle. Many times I have observed students asking for help and teachers structuring the work for students, breaking down questions and converting them into small easy steps. In doing so they empty the work of challenge and opportunities for struggle. Students complete the work and feel good, but often learn little.

I saw a very similar teaching approach, focused on struggle, in a visit to classrooms in China, another country that scores highly in mathematics. I had been asked to visit China to give a talk at a conference and managed, as I like to 'do, to sneak away and visit some classrooms. In a number of high-school math classrooms, lessons were approximately one hour long, but at no time did I see students working on more than three questions in one hour. This contrasts strongly with a typical US high-school math classroom, where students chug through about thirty questions in an hour—about ten times more. The questions worked on in Chinese classrooms were deeper and more involved than the ones in US classrooms. Teachers would ask provocative questions, deliberately making incorrect statements that students would be challenged to argue against.

One of the lessons I watched was on a topic that is often uninspiring in US classrooms—complementary and supplementary angles. The teacher in China asked the students to define a complementary angle, and the students gave their own ideas for a definition. Often the teacher would push the students' definition to a place that made it incorrect and playfully ask, "Is this right, then?" The students would groan and try to make the definition more correct. The teacher bantered with the students, playfully extending and sometimes twisting their ideas to push the students to deeper thinking. The students probed, extended, clarified, and justified for a long time, reaching depths that were impressive.

Contrast this with the standard US lesson on the same topic. Teachers often give definitions of complementary and supplementary angles to students, who then practice with thirty short questions. The defining characteristic of the lesson in China was struggle—the teacher deliberately put the students in situations where they became stuck and had to think hard. The lesson was entirely consistent with researchers' description of targeted, mistake-focused practice. As Coyle says, the best way to build a highly effective circuit is to "fire it, attend to mistakes, then fire it again." This is what the teachers in China were enabling their students to do.

Elizabeth and Robert Bjork are scientists at UCLA who have studied learning for decades. They point out that a lot of learning that happens is very unproductive, as the most important learning events often go against intuition and deviate from standard practices in schools. They highlight the importance of "desirable difficulties," again suggesting that the brain needs to be pushed to do things that are difficult. They particularly highlight the act of retrieving information from the brain, as every time we retrieve something, it changes in the brain and is more accessible when needed later.[11]

Many people study for tests by rereading materials, but the Bjorks point out that this is not very helpful for the brain. A much more helpful way of reviewing material is to test yourself, so that you keep having to recall the material—and hopefully make mistakes and correct them along the way. Learning scientists point out that these tests should not be performance events, as these cause stress and reduce the learning experience. Nonevaluative self-testing or peer testing is most beneficial.[12]

Teaching the Value of Mistakes

As neuroscience becomes more established as a field, it seems that more and more evidence is revealing the value of mistakes and struggle. Good teachers have known this intuitively and impressed upon learners that mistakes are really good opportunities for learning. Unfortunately, I have found that this message is not strong enough to keep students from feeling bad when they make mistakes—often because of the performance culture in which many good teachers work. Even when the message is phrased more powerfully—that mistakes are good not only for learning, but for brain growth and connectivity—it is hard for teachers to send it in a system in which they are made to give students tests that penalize them every time they make a mistake.

This highlights the challenge of changing education—it is a complex system that has many different parts, all of which impact each other. Teachers can give the right messages to students, but then witness their messages being undermined by a practice that is imposed by their school district. This is why I encourage any teacher who learns about effective messages and teaching ideas to share them not only with their students, but with administrators and parents as well.

When teachers encourage students to make mistakes and struggle, it is incredibly freeing. New Zealand second-grade teacher Suzanne Harris began teaching in an era of procedural teaching and timed testing. When she read one of my books, she knew that what she felt was right was backed up by research and asked her principal if she could teach the "Jo Boaler way"! He agreed. Suzanne went on to make many changes, one of which was to explain the positive benefits of mistakes and struggle to her students. In my interview with Suzanne, she described how this and other messages had changed things for a young boy in her class.

Dex, a child with a designated learning difference, had to

take medication to help him get through a school day. One day in class Suzanne had given the students one of the tasks from our youcubed website called Four 4s. This is a wonderful task that presents the following challenge:

Try to make every number between 1 and 20 using four 4s and any operation.

The students had enjoyed the task and extended it to numbers beyond 20. As Dex worked on this task, he added 64 and 16. Later when he needed to add 16 and 64, he realized that the answer would be the same. In this moment Dex had discovered the commutative property—an important characteristic of number relationships. Addition and multiplication are commutative: the order in which you work does not matter—for example, you can add 18 + 5 or 5 + 18 and get the same answer. This is important to know, as other operations, such as division and subtraction, are not commutative, and the order does matter.

Suzanne recognized that Dex had stumbled across commutativity and called it the Strategy of Knowing Reversibility for the second-grade students. Over time the other students also learned about the strategy. They found a poster from the hit TV show *The X Factor* and put Dex's strategy on it, renaming it the Dex Factor. Later in the year when students were asked to share what they had learned, one of the young girls said that she had learned the Dex Factor. Another talked about how the Dex Factor principle had helped her learn the times tables. Suzanne reflects that the other students stopped seeing Dex as a "dummy" and instead saw him as a "genius."

One day the principal came into their class and challenged the students, who had learned about the value of mistakes. He said to them, "So I can just say that 5 plus 3 is 10 and then my brain grew? I can just deliberately make a mistake. Is that how it works?"

Suzanne says the students were horrified and said to him, "What? Why would you deliberately make a mistake? Who would do that? Who does that?"

He said, "Well, you just told me that if I make a mistake, my brain will grow."

They replied, "Yeah, but you didn't actually make a mistake if you deliberately did it. You deliberately knew that it wasn't a thing, so that's not a mistake. That would be just really silly!"

I am especially pleased when I hear about students standing up for the new knowledge they acquire, even when challenged by adults. I recently received an email from a teacher, Tami Sanders, about a student who was empowered by the new knowledge she had learned. Tami teaches third-grade students at an international school in Hong Kong. This is part of the email she sent me:

> Well, the quietest child in my class came up to me today. She spoke so quietly that I had to lean down to hear her. She whispered in my ear, "Ms. Sanders, I've been reading this book from your bookshelf, and I think you really need to read it. It's so good." I looked down to see what she was holding in her hands, expecting to see a simple nonfiction trade book. To my surprise, she was handing me your book, Mathematical Mindsets. I was so touched and had to share this with you.

Gisele, the student, then wrote to me recommending that I write new books containing the same ideas for five-year-olds and younger, for six-to-eight-year-olds, for nine-to-twelve-year-olds, for thirteen-to-fifteen-year-olds, and for sixteen-year-olds and up! I have not done this—yet—but I really like her ideas and her passion to spread the ideas more broadly. Gisele also sent me this drawing, capturing the moment when she told her teacher about my book!

When we taught our youcubed summer camp for middle-school students a few years ago, we told the students that we loved mistakes, they caused brain growth, and they were a really important part of learning. This caused the students to become freer in their learning and it changed their approach in many different ways. The students became more willing to share ideas, even when they did not know if they were right; they also became persistent when faced with difficult problems. The simple idea that mistakes are good for our brains proved transformative. Receiving the positive mistakes message was a significant aspect of the change in their learning pathways and their growth as learners.

One of the camp students in the class that I cotaught with Cathy Williams was named Ellie. Ellie was one of the shortest students in the class and always wore a baseball cap set at an angle. When she gathered at the board with other students in her group to sketch out ideas, she would often bounce on her toes to reach the work of the other students. But Ellie's physical size did not match her desire to learn. If I were to choose some words that characterized Ellie during our teaching sessions, they would be "determined," "obstinate," and "I am going to figure this out if it kills me."

Ellie was one of the lowest-achieving students on our pretest. She was seventy-third out of eighty-three students. She told interviewers before she came to camp that math class was usually boring and she didn't want to come to our camp. She would have preferred to stay at home and play Minecraft. And yet Ellie was present during every moment of camp, always arguing her case, pushing for complete understanding. She sometimes became extremely frustrated when she didn't quickly understand a problem, but she stayed focused and asked question after question. She would also make many mistakes, frequently pushing herself beyond them to the correct answer. An observer watching the class in camp would probably have said she was a low-achiever who was working hard to understand.

What makes Ellie's experience so compelling? Ellie was the most improved of all of the students in camp, moving from one of the lowest scores on the pretest at the start of camp to one of the highest scores eighteen days later. She started in seventy-third position but ended in an impressive thirteenth position, moving up sixty places in the process—and improving her score by 450 percent! Ellie was working in Coyle's "zone of accelerated learning"[13]—pushing at the edge of her understanding, making mistakes and correcting them, and developing her understanding at an accelerated rate.

How do we encourage this productive learning for more people? What was different about Ellie? What made her stand out from other students was not her great understanding—she was not the student always getting answers correct—but her dogged determination even in the face of failure. When I talk with teachers, they often say this sort of persistence is missing in the students they teach. One of the most common complaints I hear from teachers is that students don't want to struggle; they want to be told what to do. To the teachers it seems as though students just can't be bothered with struggling, which is probably what it looks like. The truth is, however, that when students don't want to struggle, it is because they have a fixed mindset; at some point in their lives they have been given the idea that they cannot be successful and that struggle is an indication that they are not doing well.

Many of the eighty-three students at our camp were risk averse and unwilling to persist, like those the teachers were describing. But in our camp environment, where we actively valued mistakes and struggle, they became willing to persist, even when they found work difficult. In the moments when they turned to us, looking forlorn or despondent, and said things like, "This is too hard," we would say, "These are the greatest moments of brain growth—that feeling of it being too hard is the feeling of your brain growing. Keep going. It is really important and valuable." And they would turn back to their work. By the end of the camp we saw students who were willing to struggle and keep going when questions were hard. When we asked questions of the whole class, hands would shoot up all around the room.

When I tell teachers about Ellie, they are hungry to know how they can get more of their students to take her approach. They want their students to embrace struggle and difficulty and keep going. Jennifer Schaefer is one teacher who has learned how to encourage students to struggle in her classroom.

Jennifer teaches sixth grade in Ontario, Canada. She was one of the teachers who contacted me, because learning about the new brain science had caused her teaching to go through a profound change. Like many other women I interviewed, Jennifer describes her own upbringing as one of compliance—she was rewarded for being "nice" and "neat" and not taking risks. She said she would never answer questions in class unless she was sure she was right.

As a former youth worker, Jennifer understood the importance of boosting children's self-esteem and confidence as part of her teaching. Learning about the brain science added what she called "an extra layer" to her understanding—a layer that changed the way she teaches. Jennifer knows that when her students come to her, they have already decided if they are "smart" or not. As soon as Jennifer learned about brain growth and mindset, she spent all of September and October instilling the important ideas. Jennifer reflected:

> *It was more than just building up their confidence. It was giving them information, giving them true facts about their brain. That's the different layer that I see to it. It was something so concrete that was about their learning.*
>
> *Yeah, I've always tried to build kids' self-esteem up, but this is different. Because it attaches to their learning, it isn't just about how they feel good with their friends; it's about their learning.*

Jennifer is unusual in the way she has understood mindset and brain research and used it in her teaching and parenting. Many teachers tell their students about the important mindset information—but they don't, as Jennifer has, allow the new brain science to fully permeate and inform their teaching. I have learned over recent years that teachers' and parents' interactions with students as they work, especially

The Steps of Struggle

Which step have you reached today?

in those very important moments of struggle, are critical to the development of a growth mindset. When Jennifer first started to encourage struggle with students, she used the metaphor of steps. She now places the image of the steps on Post-it notes all around her classroom.

Jennifer tells her students that these are the steps of struggle, and they don't have to be the "smug person" on the top step, but they should also not be the "sad person" on the bottom step. They simply need to get onto the steps of struggle, as that is a really important place to be. As Jennifer described it to me:

> Nobody wants to be that guy on the bottom step. He looks kind of huffy and mad. And the guy at the top is annoying 'cause he's finished and he's super happy. I always say, "You don't have to be that guy. That guy is annoying. Be in the middle. Right?"

Jennifer's students liked the step analogy, but they liked another one even more. It comes from James Nottingham, an

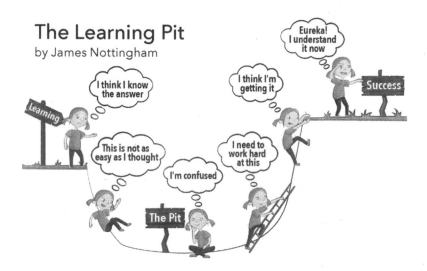

The Learning Pit
by James Nottingham

educator in the UK. He came up with the idea of the "learning pit"—the pit of struggle, a really important place to be.

Jennifer asked her students to make their own drawing of the "learning pit" as a class, and this is what they made together.

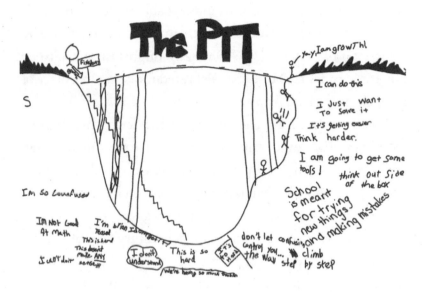

The students write about their feelings when they struggle—"I'm so confused," "I'm not good at math," "This doesn't make any sense"—feelings we have all experienced. They also write about their feelings as they progress in the struggle pit—"Don't let your confusion control you," "Climb the wall step by step," and other positive thoughts. Jennifer celebrates their being in the pit, telling them she could grab their hand and they could jump over the pit together, so they miss it altogether, but that that would not be helpful for their learning and brain growth.

She told me that students sometimes get frustrated and say to her: "Ms. Schaefer, I am really in the pit!"

And she answers, "Excellent! What classroom tools do you need?"

There are two important features of this answer. First, Jennifer celebrates the fact that the student is in the pit, saying, "Excellent!" Second, she does not structure the work for them, breaking it down into manageable pieces; she simply asks what resources would help them. This is a teacher who understands that struggle is a really important place to be and that students should be celebrated for being in a place of struggle, not saved from it.

Leah Haworth, another teacher I interviewed, highlighted a different and equally important way of attending to struggle. She related her own story and went on to describe her teaching method, which had a big impact on a young student of hers who started the school year in tears.

When Leah was a student in elementary school herself, she had some really negative experiences. She started her schooling in England and was unfortunately one of the many students put into a "low-ability group" and given the message that she was not worth bothering with. Not surprisingly, Leah developed low self-esteem but was fortunate to move to New Zealand when she was thirteen and learn from teachers who dedicated themselves to helping her catch up. When Leah became a teacher, she was already acutely aware of the importance of encouraging students—particularly those who had given up on themselves. One of those students was Kelly. When Leah saw Kelly cry, she knew it was because she had developed the idea that she was not a good student. She also knew that Kelly would become especially anxious when she struggled with work. So Leah decided to assist Kelly when work was difficult and then gradually withdraw the help. She also shared with Kelly and all the other students that when she was in school, she did not feel good about herself and she too would cry when she felt she could not do something. As the school year went on and Leah gave Kelly more and more difficult work with less and less support, Kelly changed. She became more confident, stopped crying, and started smiling. As Leah reflected:

> It was amazing to see in just over a year how far this girl had come, not only in her mathematical abilities, but her complete mindset shift and how she applied that way of thinking to other aspects of learning. She went from absolutely no self-belief that she could try any type of challenging math problem (crying and no willingness to be involved in discussion) to having a go at anything and even sharing her misconceptions openly with the class. The whole class went on that journey, but her journey was just amazing to see, and it was one of those moments that made me sit back and think, this is why I became a teacher.

Leah had shifted her teaching to encourage her students to have a growth mindset and to embrace struggle, and this made a big difference for a lot of students. Before Leah made these changes, 65 percent of her students would reach the content standards. After the first year with the new approach, 84 percent of students reached the standard. This sort of shift is typical for teachers we work with, and although we aim to get 100 percent of students to reach the standards, reaching 84 percent in the first year is impressive. Leah achieved this through giving positive messages to all of her students, including those who lacked self-belief, and encouraging important times of struggle.

When I initially brought the positive new research about mistakes and struggle to students and the public, I described it as "Mistakes grow your brain." This message is simple and powerful, and I know many students across the world have been helped by it. I have been criticized for this message by people who take "grow" literally (and narrowly) to mean that brains get bigger and bigger. What we know is that when mistakes are made, connectivity increases and brains grow in capacity and strength. I stand by my initial words because the message needs to work for students of a very young age—kindergarten and younger—and growth can take all sorts of different forms. It seems to me that greater connectivity and future capacity are very important forms of growth.

Changing Your Perspective on Failure

Learning about the positive benefits of mistakes provides a different perspective on failure. This is an important part of becoming unlocked and living a limitless life. I myself have made a transition from being locked—fearing failure and doubting myself—to being unlocked. It is a process that has to be worked on continually.

As an academic, I experience a lot of failure. To keep our youcubed center at Stanford running, supporting staff salaries and providing free materials for teachers and parents, we have to apply for lots of grants—most of which are rejected. I also have to submit our papers to journals, where rejection is part of the process. If they are not rejected, they are subject to reviewers' comments. I have had reviewers dismiss my work entirely, saying that it is "not research, just a story." It is nearly impossible to keep going as an academic without viewing "failure" as an opportunity to improve. A wise professor named Paul Black, my PhD advisor, once said to me: "Whenever you send a paper to a journal, have in mind the next journal you will send it to when the paper is rejected." I have used his advice a number of times.

Taking a limitless approach—particularly when embracing challenge and struggle—also helps when we encounter difficult people. In today's world of social media, it seems impossible to make a statement about anything without getting pushback, some of it aggressive. I have experienced extreme and aggressive pushback many times, and I now know that it is important to stay strong in those moments and to look for something positive. Instead of dismissing a challenge or beating yourself up, think, "I will take something from this situation and use it to improve."

Karen Gauthier learned to take this approach to failure after learning the new science of brain growth. Karen is a teacher and parent who grew up as a "selective mute," as she describes it, because "it was easier to not say anything than to be wrong." When Karen was a child, her parents let her give up on things that were hard. She quit softball, piano, and other activities that she found challenging. As a teacher or parent, we may think it best to let children give up, to save them from struggle. That can seem supportive, but it may backfire.

I remember being in a very effective high-school classroom at Railside, watching a student work on some mathematics at the whiteboard. She was explaining her work to the class, and then she paused. She then faltered completely, saying she didn't know what to do. The whole class was watching, and the room fell silent. This may seem to an observer like a horrible situation. But the teacher told the girl not to sit down, to keep going. The girl stayed at the front, worked through her confusion, and continued on the problem.

Later the girl reflected on that moment and said something that surprised me. She said, "The teacher didn't give up on me." Other students in the class agreed with her, interpreting their teacher's pushing them to do hard things as a sign that the teacher believed in them. This was one of the first times I made the link between pushing students or one's own children to hard places and their interpreting this as an inspiring vote of confidence.

Karen had chosen to become a teacher because she wanted to give students a better experience than she had known. An amazing teacher, Karen received the Orange County "Teacher of the Year" award. Soon after that Karen was invited to become an Orange County math coach, and it was during this time that Karen experienced what she described as complete failure. Karen had just started working as a coach and had asked teachers to try new methods—but they had not been receptive to her ideas. Karen recalled being about ten weeks into the work and realizing she was failing in her new role. And that's when she reverted back to her childhood, thinking, "Oh my gosh, I'm not good enough. Who am I fooling? I can't do this."

Karen described going through some hard days, during which time a friend helped her develop resilience and self-belief. It was during this time of self-doubt that Karen read with interest the research on mistakes and brain growth. This changed her. In Karen's words:

And then all of a sudden, I had a whole different mindset. It was like, "Wait a minute. This is an opportunity. Not something that's going to . . . where I'm gonna walk away and say, 'This is it. I'm done.'"

This is exactly the reaction that characterizes being limitless in times of challenge—thinking that the challenge is not something that will get the better of you, that it is an opportunity.

Part of Karen's journey involved realizing that failure happens to everyone—even though some people go through life acting as though it does not—and that other people also felt the way she did. Now Karen can look back on that difficult time and approach failure and struggle differently. She talked about a metaphor of a valley and a mountain, which allowed her to see herself in a new light:

When you're in the valley, you're in the deep dark trenches of change—honor that time, work through it, and someday you'll be on the mountain looking back and be grateful.

As part of her transition, Karen also talked about changing her own negative self-talk into positive self-talk and being careful to think good thoughts.

Before Karen made her transition and became unlocked, she told me that her parenting of her children mirrored that of her own childhood. She would let her children give up on activities that were hard. That has now changed, and Karen recalled a parenting experience that "closed the circle" for her:

One of the best examples was with my son. It was two years ago, when I was driving him to his last baseball game in Little League. He had never hit a home run, and as I was driving him there, he said, "Well, it's my last game. Guess I'm never gonna hit a home run."

And I was like, "Well, what do you believe? Do you believe you can?"

He was like, "I don't know."

And I said, "When you get up to that plate, you say, 'I am, I am . . .' and you fill in the blank, whatever that is. 'I am strong enough. I am good enough. I am going to hit a home run.'"

And Jo, he did. He did! 'Cause as he walked to the plate, I just shouted out, "I am." He looked back at me like, "Oh, Mom, be quiet!" and then he got up and hit that home run, and I was screaming!

Karen's process of becoming unlocked was important for her as a math coach and as a parent. Recently Karen embraced her growth mindset and applied for an even more senior post, which she was given; she is now the curriculum specialist for one of the largest districts in California. Karen said that she would never have applied for the position before learning about the importance of embracing challenge. Karen's process of becoming unlocked took a few years, and it started with learning the brain science showing the importance of struggle and mistakes. We are all on a journey in our own process of becoming limitless and being able to fully embrace challenge. Karen reached the point of seeing failure as an opportunity.

It is in the handling of failure that the quality of being limitless really shines. Those with a growth mindset may approach hard challenges well and many times succeed, but what does a growth mindset say when we fail? Those who fail and continue on undeterred, those who get knocked down and get right back up again, those who get pushback and see it as a positive sign that they are doing something important are the people who are truly limitless. It is easy to feel open and free when things are going well; it is when things are going badly and challenges or aggression stand in our way that it is most important to be limitless.

One person who exemplified this for me was Kate Rizzi. When Kate was growing up, she was extremely curious. But her family did not value curiosity; they valued following instructions and being compliant. Feeling that her curiosity was inappropriate made Kate question who she was, and as a result she did not value herself. Such disapproval caused Kate to feel as though she had to shrink herself, to make herself into a different person from the one she was. It is a devastating feeling, and one felt by many gay and transgender youth as well. This "shrinking" resulted in Kate's lack of self-confidence. Kate described the feeling in much of her childhood of "trying to prove you were smart on top of feeling not smart" and of going through school and college worrying she would be "found out."

Kate's turning point was a Landmark Education course, which taught her about the brain and about different ways to approach life. Kate learned about the limbic area of the brain—an area that developed in prehistoric times to protect early people from dangers such as saber-toothed tigers. We no longer have to worry about turning a corner and being savagely attacked by a wild animal, but the limbic part of our brain still kicks in with messages like, "Don't do that. Don't take that chance. Don't take that risk." Kate learned in her course that we can and should resist such thinking.

Kate recalled the way the course helped her be more aware of her own feelings and her ability to change her experiences. Prior to taking the course Kate said that she had always felt on "high alert" in case someone found her out, a feeling that many people have shared with me. After taking the course Kate started a process of becoming limitless that she phrased as "trying her life as an experiment."

Immediately after returning from the course, Kate saw a job advertised that she said she would never have previously considered. The job was the assistant to the dean of the school of communications at her local university. With her new approach, she decided to apply for the job, and she got it. Kate describes the experience of getting this job as her first "data" in her new "life experiment"—you can try risky things and they can work out. As time went on Kate stopped being afraid of life and started to follow her passion, guided by her interests instead of fear.

Recently, however, Kate came across a huge stumbling block. On her career path she had progressed to a position as a learning specialist, but she had only been there for four months when she was unexpectedly and inexplicably fired. The school where she was working was apparently not ready for the cutting-edge and important ideas she was promoting.

Many people would crumble in this situation, but Kate had, over many years, become unlocked and was able to re-frame her situation, seeing it not as a failure but as an opportunity. After the initial shock, Kate decided to regard her job ending as a new chance for renewal and creativity. Instead of looking for another job, Kate created her own and now works as an education facilitator connecting schools and families. Kate meets with teachers on behalf of students and says she "really loves" what she is doing now. Talking to Kate, I found it hard not to be struck by the change she has undergone—from that child and young adult who went through life "afraid of being found out" to the strong woman she is now, who does not let anything deter her.

We now know that our brains grow and change all the time. We also just learned that making mistakes and struggling leads to improved learning and growth. Together, these two keys liberate us from the fixed-brain myths that permeate so much of the world we live in. When people realize they can learn anything and that struggle is a sign of something positive, they learn differently, more positively, and also interact differently. Instead of thinking that they have to know everything, people become open to being vulnerable and to sharing uncertainty. This helps them contribute ideas in meetings instead of worrying that they will be found out for "not knowing everything." This change is freeing and liberating. The new science of brain change and the benefits of challenges can be transformative no matter whether we are students, educators, parents, or managers. The process of becoming limitless is available to all of us. And the next chapter presents a surprising and invaluable piece of science about this process.

3

CHANGING YOUR MIND, CHANGING YOUR REALITY

ALL OF THE learning keys are important, but some are really surprising. The learning key in this chapter is perhaps the most surprising of all. Put simply, when we believe different things about ourselves, our brains—and our bodies—function differently. Before I talk about the ways students transform when they hold different beliefs about themselves, I would like to share the stunning evidence of the changes that take place within our bodies—changes to our muscles and internal organs—when our perceptions of ourselves change.

LEARNING KEY #3

When we change our beliefs, our bodies and our brains physically change as well.

Beliefs and Health

In order to study the impact of our beliefs on our health, Stanford researchers Alia Crum and Octavia Zahrt collected data from 61,141 people over an extensive time span, twenty-one years. The researchers found that those people who thought they were doing more exercise were actually healthier than those who thought they were doing less, even when the amount of exercise they were doing was the same. The difference between the negative thinkers and the positive thinkers was incredible: negative thinkers were 71 percent more likely to die in the follow-up period than those who thought positively about their exercise.[1]

In another longitudinal study, researchers surveyed adults who were around the age of fifty to see how they felt about aging. The adults were then put into different groups depending on whether they had positive or negative beliefs. The positive-thinking adults lived an average of seven and a half years longer than those with negative beliefs, and these advantages remained after adjusting for baseline health and other variables.[2] In another study, of 440 participants, age eighteen to forty-nine, those who held negative ideas about aging at the beginning of the study were significantly more likely to experience a cardiovascular event during the next thirty-eight years.[3] In a study with younger adults, age eighteen to thirty-nine, those with negative beliefs were twice as likely to have a cardiac event after age sixty.[4]

In another study, Alia Crum and Ellen Langer performed an interesting experiment with hotel cleaning staff, whom they divided into two groups. One group was told that their cleaning work satisfied the US Surgeon General's recommendations for an active lifestyle. The other group was not told this information. The behavior of people in the two groups did not change, but four weeks later the group who believed their work was healthier, when compared to the control group, showed a decrease in weight, blood pressure, body fat, waist-to-hip ratio, and body mass index! This result reveals that what we think about our exercise can cause our weight to drop and our health to improve.[5] The researchers found that, prior to the beginning of the study, the workers did not think they were exercising well, and so learning that their work was good exercise had a big impact on them. This belief changed their mindset about their exercise and—probably—their lives. The improvement in mindset changed the functioning of their bodies, just as we know that a change in mindset affects the functioning of the brain.

New studies also show that with mental focus we can develop our muscle strength and become more accomplished in learning to play musical instruments—without even practicing or working out. Researchers trained people to develop muscle strength without using their muscles. Instead, they were just thinking about using them.[6] The participants in the study engaged in either mental training or physical training. In the mental training participants were asked to imagine their finger pushing hard against something. In the physical training participants were asked to actually push with their finger to develop muscle strength. The training lasted for twelve weeks, with five fifteen-minute trainings each week. The group imagining pushing increased their physical strength by 35 percent. The physical group increased their strength by 53 percent.

The researchers explain this result, that strength improved without the muscles actually moving, by saying that the mental training improved the cortical output signal, which drives the muscles to a higher activation level and increases strength. The study concludes, "The mind has remarkable power over the body and its muscles." When I told my colleagues about this study, they joked that they were so pleased they no longer needed to go to the gym; all they had to do was think about going! But they are, in part, correct. When our mind imagines in a focused way the development of muscles, the muscles actually strengthen through the development of enhanced signaling in the brain.

A similarly impressive result came from a study of pianists.[7] Professional pianists were recruited to learn and perform a musical piece, but half of the pianists trained by imagining playing the music and the other half trained by actually playing it. The group who imagined playing the piece not only improved their performance so that they were almost indistinguishable from those who actually played it, but they improved in all of the same ways as the actual players, in movement velocity, movement timing, and movement-anticipation patterns. Scholars point out that imaginary practice is beneficial for pianists, because it can save them from overuse of their hands, which may lead to physical strain.[8]

Beliefs and Conflict

It is no exaggeration to say that my colleague Carol Dweck has changed the lives of millions of people through her work on mindsets. We all have different mindsets about our own ability. Some people believe they can learn anything; others believe their intelligence is fixed and there are limits to what they can learn. She and her teams have conducted many different studies showing that the ideas we hold about ourselves matter. Before I share some of the ways that changing ideas can change one's achievement, I will highlight some incredible new work that the mindset team has been conducting on people's ability to move out of conflict and become more peaceful.

I first met David Yeager when he was a doctoral student at Stanford; he is now a psychologist at the University of Texas at Austin. He and Carol Dweck have conducted important research on mindsets and conflict. They found that people with fixed mindsets (individuals who believe that their abilities and qualities are static and can't be changed) have a heightened drive for aggressive retaliation during conflicts. Yet when they are given information that causes them to change their mindset, these aggressive tendencies fade away.[9]

The researchers point out that people with a fixed mindset are more likely to be aggressive, since they believe that people—including themselves—cannot change and that any failure they have experienced themselves is an indication of their own weakness. This causes them to harbor more negative feelings about themselves. Those with a fixed mindset feel more shame, view their adversaries as bad people, and express hatred toward them.

The researchers found that people with a growth mindset respond to conflict with less hatred, less shame, and less desire to be aggressive. Their improved response to conflict comes about because they view others as being capable of change. Importantly, the aggressive feelings that those with a fixed mindset experienced were not fixed, and when they were helped to develop a growth mindset, they became more forgiving and wanted to help people act in better ways in the future.

In other studies, researchers have found that people with

growth mindsets are less prejudiced about race.[10] When people learn that others are not fixed in their thinking and could become less prejudiced, they change the ways they interact with those from other racial groups.

These new studies say something important about the ways fixed beliefs impact many aspects of our lives. The research also reveals that when people change their mindset and believe in personal change, they open up in different ways, including feeling less aggression toward others. Not only that, but science shows us that changing our beliefs improves our health and well-being. Given these impressive outcomes, it is perhaps not surprising that when we change our beliefs about our own learning and potential, our achievement significantly improves.

Beliefs and Learning

A landmark study by Lisa Blackwell, Kali Trzesniewski, and Carol Dweck clearly showed the impact of different beliefs on students' learning. The study involved seventh- and eighth-grade math students.[11] The students were divided into two groups that differed only in the ways they thought about themselves—their mindsets. The two groups of students went to the same school and had the same teachers. The graph below shows that students who held positive beliefs were on an upward trajectory in their achievement, but those with a fixed mindset stayed constant and were lower-achieving. There have now been many studies that have replicated this result, revealing the importance of people's mindsets—at any age.

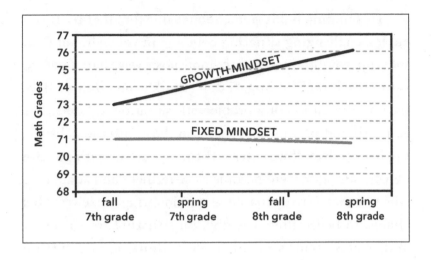

In the last chapter, I reported a study conducted by Jason Moser and his colleagues showing that mistakes are productive for brain activity and growth.[12] This study also highlighted that people with a growth mindset experienced significantly greater brain activity when they made mistakes than people with a fixed mindset did. The team created voltage maps, maps showing activity or heat, in the brains in the study. The voltage maps showed that the brains of people with a growth mindset were glowing orange—almost as though they were on fire with growth compared to those of people with a fixed mindset.

This research result reveals something extremely important—it shows in concrete terms that what you believe about yourself actually changes how your brain operates. For many years people have believed that our emotions are separate from our cognition or knowledge, but this is not the case. They are in fact intertwined. When making mistakes, those who believed in their own potential experienced more beneficial brain activity than those who did not believe in their own potential.

This finding holds tremendous value for all of us. If you enter a challenging situation believing in yourself, but then mess up, your brain will react more positively than if you go into a situation thinking, "I don't think I can do this." If we have a difficult job or a problematic situation at home, this result should prompt us to go into those situations believing in ourselves. If we enter difficult situations with positive beliefs, our brains will become more resilient and adaptive when we make errors than if we are doubting ourselves. This change in belief alters the physical structures of the brain and creates avenues for higher-level thinking and creative problem solving. Just as those who believed they were engaging in healthy exercise became healthier, those who believe they are learning more productively actually learn more.

The result of Moser's study also helps us understand the graph from the Blackwell, Trzesniewski, and Dweck study, which shows an upward achievement trajectory for students with a growth mindset. This trajectory seems less surprising when we know that the students with a growth mindset were experiencing more brain activity every time they made a mistake. This single research result sheds light on the low achievement of students in schools—as so many students believe they are not cut out for subjects they are learning. We now know that the idea that some people are "math people" and some are not is a harmful myth.

Changing Mindsets

Considerable evidence exists showing the relevance of the progress that can be made when students believe in their learning potential and let go of ideas that their achievement is genetically determined. Therefore, it is critical to create opportunities for our students, our children, and people we work with to develop a growth mindset and to understand where different mindsets come from. One of Dweck's studies showed that children's mindsets, which had come from the type of praise given by parents, were developed by the time they were three years old. In their study, Dweck and her colleagues found that the praise given to children fourteen to thirty-eight months old predicted the mindsets they had when the children were seven to eight.[13] The damaging praise given by parents was the kind that instills the idea of fixed ability. When children are told they are smart, they at first think that is good, but when they mess up on something, they decide they are not smart, and they keep evaluating themselves against that fixed idea.

One of Dweck's studies revealed the immediate impact of the word "smart." Two groups of students were given a challenging task. On completion, one group was praised for being "really smart," and the other was praised for working hard. Both groups were then offered a choice between two follow-up tasks, one that was described as easy and one that was described as challenging. Ninety percent of the students praised for working hard chose the harder task, whereas the majority of the students praised for being "smart" chose the easy task.[14] When students are praised for being smart, they want to keep the label; they choose an easy follow-up task, so they can continue to look "smart."

This type of thinking is often at play when students choose to opt out of challenging courses, such as math and science. The largest group of fixed-mindset students in the

US school system is high-achieving girls. In another of their studies, Dweck and her colleagues found that the girls most likely to drop out of math and science were those with a fixed mindset. One of the studies was conducted at Columbia's math department, where researchers found stereotyping to be alive and well—women were being given the message they did not belong. Women with a growth mindset were able to reject such messages and continue on; those with a fixed mindset dropped out of their STEM classes.[15]

So how do we develop a growth mindset? The process is a journey, and not a switch that can be flipped causing instant change. But mindsets can change. In a number of studies, people's mindsets have changed when they have been shown the evidence of brain growth and plasticity you have read about in the last two chapters. This has also been my own experience in the classroom and in the workshops I conduct. Once people become aware of the science, growth and change starts to take place. I see it in students and hear about it from teachers all around the globe. The science also supports this.

When students receive the information that the brain is like a muscle that grows with effort and work, their levels of achievement change. In an important mindset study, a group of researchers created an experiment in which work-shops were offered to two groups of seventh graders; one group received information on study skills and the other group got material on brain growth and mindset.[16] The students' overall achievement levels were declining over the course of seventh grade, but for the students who received a mindset workshop, the decline was reversed and their levels of achievement improved.

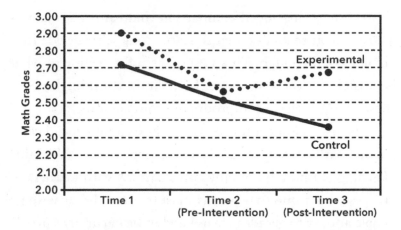

Part of the reason Cathy Williams and I founded you-cubed, our Stanford center, was to share evidence of brain growth and mindset with students. In the interview study we conducted for this book with sixty-two adults who reported change, it became clear that people can change at any age. The interviews also revealed the detailed ways that fixed ideas can lock people up and, conversely, the ways that mind-set and growth ideas can unlock them.

Mariève Gagne, a teacher in Canada, is a native French speaker who grew up, like many others, thinking that she was not good enough for STEM subjects. Interestingly, she developed this damaging belief even though she was in the highest ability group at her school. The reason she felt she was not good enough was that she was not one of the top students in her group. This tells us the extent of the damaged thinking. Even students in advanced groups believe they are not good enough if they are not at the top of their group.

In our youcubed film about the damage of the "gifted" label, one of the Stanford undergraduates, Jodie, says that she believed she could not continue to study engineering because she was "not the best student in her math or chemistry classes." It is this social comparison that causes students—who arrive at school excited to learn—to quickly decide they are not good enough. This is the start of a decline in what is possible for these students. Fixed-brain thinking leads to these rigid beliefs and unproductive comparisons.

It is just as important to take on ideas about social comparison with students as it is to make them aware of the value of struggle. I have had many conversations with learners of all ages who argue that brains must be fixed, because some people appear to get ideas faster and to be naturally "gifted" at certain subject areas. What they do not realize is that brains are growing and changing every day. Every moment is an opportunity for brain growth and development. Some have simply developed stronger pathways on a different time line. It is critical that students understand that they too can develop those pathways at any time—they can catch up with other students if they take the right approach to learning.

Angela Duckworth, the author of the bestselling book *Grit*,[17] highlights this point with her recollection of David, an algebra student she taught as part of a low-track class at Lowell High School in San Francisco. Low-track students, of course, would not get access to higher-level courses. David worked really hard and did well on all of her assessments, so she arranged for him to move into the higher track. David's transition was bumpy, and he at first got Ds on tests in his new class. But David approached this as an opportunity to work out what he was doing wrong and improve.

By his senior year David was taking the harder of the two versions of calculus and achieved 5 out of 5 in the AP exam. David went on to attend Swarthmore and study engineering and is now an astronautical engineer. Things could have been so different for David if he had given up when placed into low-track math as a freshman and if he had not had a teacher who fought for him to move to the higher track. There are many other students who do not get such opportunities and who end up as low-achievers simply because they did not get access to higher-level content or because they did not believe they could be successful.

When students become disillusioned because others are ahead of them or complain that they don't understand something, a word that Carol Dweck champions using with them is "yet." When I ask adults to visually represent an idea, I often hear them say, "I am terrible at drawing." I tell them, "You mean you have not learned to draw well yet." This may seem like a small linguistic change, but it is an important one. It moves the focus from the perceived personal lack to the process of learning.

It is similarly important for teachers to start the first class of the year by sharing the science of brain growth and telling students they may not be the same as each other, but anyone can learn the content that is being taught and productive learning is in part due to their thinking. This message is liberating and the opposite of that teacher message I mentioned earlier: "Only a few of you will be successful in this class." In Chapter 1, I noted the research on the lower proportions of women and students of color in the fields where professors believe that students need a gift to be successful. This result comes about, in part, because the teachers and professors who believe in giftedness communicate to students that only some students will be successful. And when they communicate that idea, only some students *are* successful.

We parents have many opportunities to observe the damage of social comparison and to talk to our own children in ways that can help ameliorate it. Siblings have millions of opportunities to compare themselves to each other, and many children develop negative ideas about their potential because they think a sibling can learn things more easily. Social comparison is particularly damaging when it is based on supposed genetic endowment. When children think that their sibling or classmate was born with a better brain and that brain will always be superior, it is demoralizing. It would be better if, instead, children saw a peer's or sibling's ability as a challenge and an opportunity—"Because they can do it, I can too."

When students learn about brain growth and mindset, they realize something critically important—no matter where they are in their learning, they can improve and eventually excel. This was demonstrated in a study of students entering high school. Sixty-eight percent of the students experienced a drop in grades in their first semester and reported feeling stressed as a result (telling us something about the damage of grading practices).[18] But students who had a growth mindset were more likely to view the setback as temporary and had lower stress levels. Students with a fixed mindset perceived the setbacks more negatively and had higher stress levels.[19] This makes sense, as students with a fixed mindset view any times of low achievement as evidence that they do not have the right kind of brain.

Mariève, who had felt unable to pursue STEM courses because she was not the highest student in her group, developed a limitless mind as an adult when she learned about neuroplasticity. She then found like-minded communities of teachers on social media who shared positive beliefs in students and knowledge of brain growth. When she joined Twitter, she was shocked to find so much good information available. In her interview with me, she remarked, "Wow, where have I been all these years?"

She became so excited about students' potential, and her own, that she has got herself a math tattoo and now teaches high-school math to adults who did not get a diploma—people who stand to benefit immensely from the scientific information and encouragement she can give them. Talking to Mariève and hearing her excitement about learning made me reflect that this amazing teacher might never have found her way back to mathematics if she had not read about neuroscience and realized that the perceptions she had lived with—that she did not have the "right brain"—were wrong.

Part of the process of change and of becoming limitless involves letting go of the idea that your past failures came about because there was something wrong with you. A similarly important change is realizing that you do not have to live your life as an "expert," that you can go into situations and proudly share uncertainty. Jesse Melgares told me about these two aspects of the change he went through as he became unlocked. Jesse is an assistant principal in east LA, but in earlier years he taught mathematics and was, as he said, "extremely self-conscious," thinking he did not know enough and nothing could change. When Jesse became an assistant principal, he needed to coach math teachers, but he was fearful that others would find out he was a fraud:

> To be honest I would get a lot of paralyzing stress when someone asked me a math-related question. . . . It was terrible. It was like a boot on my chest. It's what I woke up with

in the morning, wondering, "Am I gonna be asked something
that I don't know the answer to? And will I be discovered as
some sort of fraud?"

The feeling of paralyzing stress Jesse described, the fear
of being asked something he could not answer, is a feeling
shared by millions of people in different situations and jobs,
and it is a feeling that I hope this book can change. For Jesse
the change began when he took one of my online courses and
realized: "Everything that I had been taught as a student of
math when I was in the K–12 system and as a math educator
was wrong."

For Jesse, the first step in becoming unlocked was realizing that any trouble he had had learning in the past was
not due to some deficit in him, but to the faulty system in
place. This is a shift I have seen others make, and it is vital for
those who have had bad learning experiences. Our summer-
camp students who were underachieving before they came
to us also had this experience—they had been thinking they
had trouble with mathematics because there was something
wrong with them. They came to realize that their lack of
achievement was due to the problems in the educational
system. This allowed them to start a new relationship with
mathematics. The same realization allowed Jesse to change
as a person.

Jesse not only started feeling better about mathematics;
he started a new "journey" discovering that mathematics
was his passion. He shifted from feeling defeatist about
math to seeing it as an exciting challenge. Jesse is now the
director of mathematics for twenty-five schools—quite a
change for a person who used to feel paralyzing stress when

he thought about math. New knowledge about the brain allowed him to shift his perspective, his mindset, and his belief in himself. Jesse still meets questions he cannot answer, but instead of being afraid, he thinks: "Well, I don't know what the answers are but, you know, we'll figure it out. This is a challenge." This shift in perspective is typical for people who have become unlocked. When people change their mindset and become aware of the positive benefits of struggle, they take a new and much more positive approach to challenge and uncertainty. They let go of the need to be the expert and replace it with curiosity and the desire to collaborate.

One of the obstacles to a positive change in our beliefs is our own self-doubt. Swedish psychologist Anders Ericsson is helpful in pointing out that self-doubt, especially when we cannot see how to move forward, is a natural part of our lives. What is not natural is a "true dead-stop obstacle, one that is impossible to get around, over, or through."[20] In all of his years of research studies, Ericsson has found it surprisingly rare to find any real limit on performance—instead, he sees people become limited because they give up and stop trying.

Recently I was watching my favorite TV show, *Madam Secretary*, a fictional depiction of the activities of the US Secretary of State, played by Téa Leoni, and her advisory staff. I find the show fascinating in its portrayal of world events, but what really draws me to it is the positive problem-solving mindset of the lead character. The episode I was watching showed a fictional crisis in West Africa that looked like it was going to lead to the death of a group of people called the Beko. Madam Secretary and her staff were trying to find ways to stop the imminent genocide. After a hard day of unsuccessful efforts, Jay, the chief policy advisor, turned to Madam Secretary and said, "We are out of options."

This negative (though understandable) statement would have prompted many people to agree, slump down in their chair, and give up. Instead, Madam Secretary looked Jay squarely in the eye and said, "I refuse to accept that, Jay." Her response gave the team inspiration, and they went on to find a creative solution to the crisis. When I watch the ways that the positive words and mindset of Madam Secretary inspire her staff, even though the show is fictional(!), it reminds me how important it is for leaders to model a growth mindset for the people they work with.

In one of my interviews, I heard a moving recollection of a belief message given by a manager in a fruit field in California's Central Valley. The message given by the manager created a new life for a boy who was working there, and that boy went on to change many other lives through his work. Daniel Rocha was the boy in the fruit field, and he is now a curriculum coach in the Central Valley. Daniel might never have held this important and prestigious job if not for the words of a farmer in a fruit field during the summer before his senior year in high school.

Daniel told me that his father was a farm laborer and that he spent his summer and winter vacation days working with his father in the fruit fields while other students were on vacation. The summer before his senior year, Daniel wanted a pair of Jordan sneakers. He planned on using the money he would make from working with his dad. The work that summer was much harder than Daniel had expected or previously experienced, and he quickly realized the sneakers were not worth the effort. But the most noteworthy part of Daniel's difficult, backbreaking summer was the message that changed his life:

We were working in a field when I noticed the owner of the orchard, the farmer, coming. Since my dad was the foreman, the farmer came over and started talking to my dad. He said, "Hey, Rocha, who is that? Who's the boy?"

And my dad in his broken English said, "That's my son."

"Well, does your son have papers?" the farmer asked.

"Yes. Of course he does," my dad said.

The farmer mumbled something to my dad, but I didn't want to look that way. I didn't want to draw any attention to myself. But the next thing I knew, I was standing on the top of a ladder holding about a forty- maybe fifty-pound sack of fruit in front of me, trying to keep my balance on the top of the ladder, when suddenly the ladder starts shaking. As I began to lose my balance, I looked down to see that it was the farmer who was shaking the ladder.

He shouted to me angrily, "What are you doing here?"

Nervously, I replied, "I'm just trying to work."

The farmer insistently shouted, "You need to get out of my fields! I don't ever want to see you here again! Let this be the last time I ever see you in these fields again. Next year you better be in college, and I better not hear that you're here anymore."

It shook me. It shook me to my core. And that day, as we were driving home, my dad turned to me and said, "¿Quieres regresar a el campo o quieres ir a la escuela?" which means, "Do you want to work in the fields or do you want to go to school?"

"Well, I want to go to school," I said.

*My dad replied with a heavy heart, "Then you got to fig-
ure that out, because I can't help you anymore. I don't know
how to, and I can't help you anymore. So just figure it out."*

*When I returned to school, I found a teacher who was
helping other students complete their financial-aid applica-
tions. I walked in and said, "I need some help." And that
brought me to where I'm at.*

The farmer shaking the ladder was a fortunate event for
Daniel, because he had not heard the idea that he should go
to college anywhere else. Recently Daniel's father visited him.
Daniel had just come from work and was still wearing a suit
and tie. Daniel's dad looked at him, said, "Look at you," and
became quite emotional, realizing how far Daniel had come
in his life. Daniel was an amazing teacher before he became
a curriculum leader and coach and made a point of communicating to all of his students that he believed in them. He
knew from personal experience just how important that message was for all of his students. Daniel not only told his students that anything was possible; he helped them believe it.

Research is telling us what some would never have believed a few years ago—that when we have positive self-beliefs
about what we can do, our brains and bodies function differently and lead to more positive outcomes. In this chapter
we have learned from research and from personal stories of
incredible change inspired by a few words. These few words
shifted perspectives—hotel workers were told that their jobs
were healthy; Daniel was told that he should go to college.
These words changed people's mindsets about their bodies
and about what they could do with their lives, which went

on to change their actual bodies and their lives. These kinds of transformations are available to us all. We can improve our own lives by thinking differently. And we can change the lives of others by encouraging them through positive thinking and knowledge of the growth and change that anyone can achieve.

Recently I chatted with Carol Dweck when we had both been asked to speak to a group of visiting Australians at Stanford. She told me that she has changed her thinking about two aspects of the way mindset functions. She had started her career thinking that people have either a growth or a fixed mindset, but she has now realized that we all have different mindsets at different times and places. We need to recognize when we are having fixed-mindset moments and even to name them.

That day, she recounted working with a business-team manager who had decided to name his fixed mindset Duane, saying, "When we're in a crunch, Duane shows up. He makes me supercritical of everyone, and I get bossy and demanding rather than supportive." A female team member responded to him, saying, "Yes, and when your Duane shows up, my Ianna comes roaring out. Ianna responds to the macho guy who makes me feel incompetent. So your Duane brings out my Ianna, and I become cowering and anxious, which infuriates Duane."[21] Carol talks about the importance of being in touch with your different mindset personas, because the more you can be on the lookout for your fixed-mindset thinking, the more you can be ready to greet it and warn it to stop.

Carol also shared her updated thinking on the dangers of "false growth mindset"[22] thinking—which involves a fundamental misinterpretation of what mindset means. She explains that "false growth mindset" thinking is telling students they simply need to try harder and praising them for effort even if they fail. She says that this backfires because students know that the praise is a consolation prize. Instead, teachers and others should praise the learning process and, if students are not making progress, help them find other strategies and different approaches. Crucially, praise should be linked to effort that leads to something important. A student might ultimately fail on a problem, but a teacher could praise the fact that correct thinking was used for part of it or that the effort led to some result that could be used to build on moving forward.

Teachers hold an incredible amount of influence. They can change the pathways of students, as many of the interviews I have shared have shown. They do this when they communicate to students that they believe in them, that they will value times of struggle and mistakes, and that they will honor different types of thinking and ways of approaching life. Parents play a similar role in valuing their children's ways of being and unlocking their children to be the people they can be.

Brains and bodies, it turns out, are incredibly adaptable. The power of this knowledge is amplified when teachers, parents, coaches, and managers, as well as students and other learners, approach learning with this in mind. A Resources section at the end of the book contains a range of free resources for parents and teachers, from videos for different age learners demonstrating important science to posters, lessons, tasks, and short articles.

We now have extensive evidence that highlights the potential of brains and bodies to change and that calls into

question the myths of "natural genius" and "giftedness."
Once we know that brains and people can achieve almost
anything, it should lead us to think of human potential—and
institutions of learning—entirely differently. But we will not
realize the potential of the new information on brain growth
and mindset without a different approach to learning drawn
from new findings in neuroscience, one that I will be sharing
in the next chapters. When we combine different approaches
to thinking about ourselves with new approaches to knowledge, the results are powerful.

4

THE CONNECTED BRAIN

IT IS SO IMPORTANT to approach life with a growth mindset, knowing that on the other end of struggle is success and that nothing is out of reach. At this point in time many people are aware of the importance of growth-mindset thinking. But there is another important part to living a life without limits that is not well known, yet is critical to keeping pathway options open. It is a different and more dynamic way to interact with the ideas you will encounter, both in any academic content you are attempting to learn and in other areas of your life.

LEARNING KEY #4

Neural pathways and learning are optimized when considering ideas with a multidimensional approach.

Although there has been a great deal of justifiable attention paid to mindset and the need to believe in ourselves at all times, when it comes to learning, telling students to have a growth mindset is not enough to overcome the conflicting messages in our culture. Carol Dweck herself has written that the information on the value of changing mindsets needs to be accompanied by a different approach to teaching, one that enables students to learn differently. One of the things that she says keeps her up at night is when students are told to put in effort and that success is all about hard work, without their being given the tools by teachers to learn more effectively. As she says, "Effort is key for students' achievement, but it is not the only thing. Students need to try new strategies and seek input from others when they are stuck."

Alfie Kohn, a great educational writer and leader, has criticized the mindset movement, saying it is unfair to tell students to change—to tell them to try harder—without changing the system.[1] I agree completely, and over many years I have learned something important: for students to develop a growth mindset, teachers need to teach with a growth perspective, opening content to the multiple ways students can learn, so that students can see the potential for growth inside it. It is challenging for students to develop a growth mindset when subjects are presented in a fixed way—as a series of questions with one answer and one method to get to it.

So how do teachers, parents, and leaders teach in ways that amplify and support the positive messages about growth and learning? The solution is a multidimensional approach to teaching and learning. This approach draws from the recent neuroscience out of Stanford and elsewhere as well as from the many different experiences of those who teach across the K–16 age range.

In my work at Stanford I collaborate with neuroscientists, in particular a group of researchers led by Vinod Menon at the medical school. Lang Chen, a neuroscientist in Menon's lab, works with youcubed on a regular basis. The researchers at Stanford study the interacting networks in the brain, particularly focusing on the ways the brain works when it is, for example, solving mathematics problems. They have found that even when we work on a simple arithmetic question, five different brain areas are involved, and two of them are visual pathways.[2] The dorsal visual pathway is the main brain region for representing quantity.

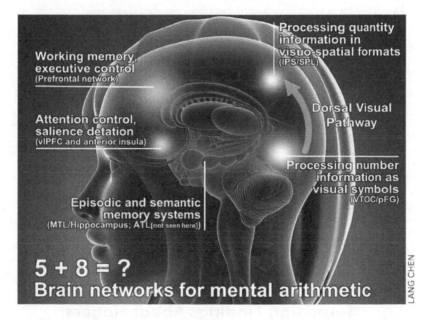

They and other neuroscientists have also found that communication between the different brain areas enhances learning and performance. In 2013 research scientists Joon-koo Park and Elizabeth Brannon reported on a study in which they found that different areas of the brain were involved when people worked with symbols, such as numbers, than when they worked with visual and spatial information, such as an array of dots.[3] The researchers also found that mathematics learning and performance were optimized when these two areas of the brain were communicating with each other. We can learn mathematical ideas through numbers, but we can also learn them through words, visuals, models, algorithms, tables, and graphs; from moving and touching; and from other representations. But when we learn by using two or more of these means and the different areas of the brain responsible for each communicate with each other, the learning experience is maximized. This has not been known until recently and has rarely been made use of in education.

The researchers who look at the interplay of different brain areas have chosen to study what happens when people work on mathematics, but the results apply to all content areas. Learning new knowledge requires different pathways in the brain—pathways that focus on attention, memory, reasoning, communication, and visualization, for example. When we stimulate all of those pathways by looking at knowledge in a multidimensional approach, our brains are strengthened, and learning is maximized.

Surprising Findings About Fingers

The new emerging details of the ways the brain processes mathematics are sometimes surprising—for example, the research showing the importance of fingers to mathematical understanding. Researchers Ilaria Berteletti and James R. Booth analyzed a specific region of the brain dedicated to the perception and representation of fingers, known as the somatosensory finger area. They found that when eight-to-thirteen-year-olds were given complex subtraction problems, the somatosensory finger area lit up, even though the students did not use their fingers.[4] Remarkably, we "see" a representation of our fingers in our brains even when we do not use fingers in a calculation. This finger-representation area was, according to their study, also engaged to a greater extent with more complex problems that involved higher numbers and more manipulation.

Because of research showing the relationships between fingers and mathematical thinking, neuroscientists highlight the importance of "finger perception"—knowing your individual fingers really well. A test for finger perception is to hide one of your hands under a book or table and ask someone to touch your fingertips. People with good finger perception can identify the fingers being touched with ease. A more challenging finger-perception test is to touch fingers in two different places—the fingertip and the mid-finger area. Here are some interesting facts about finger perception:

- The extent of college students' finger perception predicts their scores on calculation tests.[5]

- Finger perception in first grade is a better predictor of mathematics achievement in second grade than tests.[6]

- Musicians' achievement in higher mathematics, a relationship that has been noted for many years, is now thought to be due to their opportunities for developing good finger perception.[7]

Neuroscientists know that it is important for young children to develop the finger area of the brain, which comes about when they use their fingers to represent numbers. Despite the knowledge of the importance of fingers, many schools and teachers discourage finger use, and students often regard it as babyish. I have tried to help with this situation by communicating the new neuroscience widely in news, media, and journal articles. In addition, I am working now with an interdisciplinary group of neuroscientists, engineers, and educators to make small robotic devices to encourage finger perception in young children. The new discoveries about the working of the brain are revealing the need for a different approach to teaching that is more physical, multidimensional, and creative than the approaches that have been used in the past in most institutions of learning.

What About Trailblazers?

In their quest to better understand the ways people become high achieving, scientists have uncovered additional interesting data on brain communication. Some people who were really high achieving, and have made great contributions to music and science, such as Mozart, Curie, and Einstein, are often regarded in a fixed way as "geniuses." But Anders Ericsson, Daniel Coyle, and others who study expertise have shown that the great achievements of these super-high-achieving people came from extreme dedication and hard work over many years.

Ericsson takes on the idea that Mozart was born with a special gift and recalls the activities he engaged in that led to his great musical accomplishments, even at a young age. He points out that Mozart was known to have had what is often referred to as "perfect pitch." This seems to be a great example of a genetic gift, as only one in ten thousand people in "normal circumstances" has perfect pitch. But careful consideration of Mozart's upbringing shows that from the age of three he engaged in the activities that develop a perfect pitch.[8]

The Japanese psychologist Ayako Sakakibara reported on a study in which twenty-four students were taught to develop "perfect pitch." The children used colored flags to identify chords and kept working on this until they could identify all chords perfectly. In this study every single one of the students developed perfect pitch.[9] This is an example of a quality that people believe to be a "gift," but that actually comes from a particular kind of learning—the kind that engages students via multiple pathways, in this case by connecting visual ideas with sounds.

Albert Einstein, probably the person most thought of as a "genius," embraced mistakes and approached learning in a particularly productive way. Some of my favorite quotes from Einstein include:

> A person who never made a mistake never tried something new.

> It's not that I am smart. It is just that I stay with problems longer.

> I have no special talent. I am only passionately curious.

> In the middle of difficulty lies opportunity.

These and other quotes from Einstein suggest strongly that he had a growth mindset, even though mindset was not a concept at the time that he was alive. Einstein talked about embracing struggle, staying for a long time with difficult problems, being curious, and making mistakes and rejected fixed ideas of talent and giftedness.

Einstein also engaged with ideas visually. Einstein often

said that all of his thinking was visual—and he struggled to then turn his visual ideas into words and symbols.[10] Einstein had a long-lasting impact on science, and it is no wonder that people regard him as a "genius." He did not have the tools and technologies we have today, but through his own thinking he predicted that black holes orbiting each other would create ripples in the fabric of space-time. It took a hundred years and what *National Geographic* describes as "enormous computational power" to prove him right. Despite Einstein's incredible achievements, he was quick to point out they did not come from a gift or special talent but dedication and hard work and what appears to be a visual approach to knowledge. Einstein seemed to take a limitless approach to learning and life, which had a positive impact on everything he studied.

According to a recent *National Geographic* article called "What Makes a Genius?" Einstein's brain, which sits on forty-six microscope slides in a museum in Philadelphia, has been examined for special qualities.[11] Many visitors have stared at Einstein's brain and not noticed anything remarkable. A team at the Imagination Institute, led by Scott Barry Kaufman, is taking a different approach by examining the brains of living people who have achieved incredible feats. They have found something interesting. What is different in the brains of people who "are trailblazers in their fields" is that they have more active connections between different brain areas, more communication between the two hemispheres of the brain, and more flexibility in their thinking.[12] The brain communication that is a characteristic of the brains of "trailblazers" is not something they are born with; it is something they develop through learning.

Ways to Encourage Brain Communication and Development

When students in school are given worksheets with a series of almost identical questions to work through, which happens a lot in math classes, they are missing opportunities to strengthen their brains and to encourage the communication that the trailblazers displayed. A much better practice is to take a small number of the questions (three or four) and approach them in different ways. In mathematics, for example, you could approach any numerical content with questions such as these:

- Can you solve the question with numbers?

- Can you solve the question with visuals that connect to the numbers through color coding?

- Can you write a story that captures the question?

- Can you create another representation of the ideas? A sketch, doodle, physical object, or form of movement?

One of the ways we encourage this multidimensional approach is through what my colleague and fellow youcubed director Cathy Williams calls "diamond paper." This is a regular piece of paper folded this way.

First fold the paper in half	Then in half again	Then fold a triangle	Then open the paper

We encourage teachers to put a mathematics problem in the center of the diamond and to use the four quadrants to invite different forms of thinking such as those highlighted in the previous list. So instead of a division worksheet (top), a diamond paper for the problem 50 ÷ 8 (bottom) might look like this.

From this:

Division 6-12   Name:

9)81	11)121	7)21	10)10	10)10
10)50	7)49	10)50	9)27	8)64
10)90	9)63	8)96	7)77	10)90
12)36	11)11	11)11	12)132	6)30
6)54	11)55	12)84	11)55	9)45

To this:

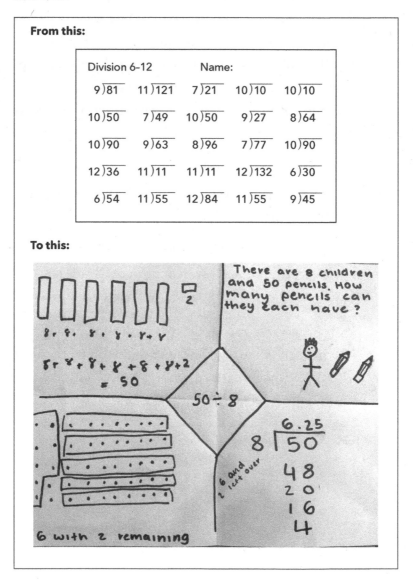

Approaching content in a multidimensional way is important in all subject areas. In English class, for example, students could study a play such as *Romeo and Juliet* by reading it and analyzing themes. Or they could take one theme—say the family—and explore it in different ways, finding a music video that captures the theme, creating their own video, writing a graphic novel, creating a PowerPoint presentation, or making a sculpture.* Such multi-modal thinking creates the opportunity for brain communication and development. Fluid and flexible brains, neuroscientists conclude, come from the synchrony that occurs when multiple brain areas are working together.[13] Communication between brain areas comes about when we approach knowledge through multiple avenues, encountering ideas in different forms and representations.

A multidimensional approach can be used in the teaching of all subjects to bring about higher engagement and achievement. Many subject areas, particularly in the humanities, already value treating the subject in multiple ways by asking students to give their own interpretations of texts they read and employing such forms as group discussions, debates, and plays. In most cases they could still become more multidimensional, but they are rarely as narrowly taught as some other subjects. In my experience the subjects that seem most in need of change are mathematics, science, and language teaching. Coming at the subject matter from multiple angles is an ideal learning approach for all of these disciplines.

For example, an innovative foreign language teacher I met with asks students to stand in a circle and tells them they are each a famous person who speaks that language. When they

---

* My thanks to Antero Garcia for his ideas about a multidimensional approach to English.

are tapped on the shoulder, students share something that person might be feeling. This is a simple but creative teaching idea that goes beyond the reading of translated words and phrases; students learn through speaking and interpreting someone else's ideas.

I will never understand the narrow teaching of science—as a list of facts and rules. That is the perfect way to turn students away from a subject that, at its heart, is about discovery, experimentation, and the possibility of multiple causes and outcomes. We need to engage students in the wonder of science, just as we need to engage them in the wonder of mathematics. This is much more important than memorizing the laws of thermodynamics (which can be looked up in a book or online).

One of my favorite approaches to science comes from John Muir Laws, a passionate nature enthusiast and educator. I love his book *The Laws Guide to Nature Drawing and Journaling*. This sounds like a book on nature, but Laws takes us through so many scientific principles in the book—and, importantly, he uses multiple lines of inquiry for his subjects. His ideas for studying nature extend to many scientific areas. He proposes that people study events by collecting data; finding patterns, exceptions, and changes over time; recording events; and making maps, cross sections, and diagrams. He then shows the multiple ways to dig into data, including writing, diagramming, recording sound, making a list, counting and measuring, using data tools, and building a "curiosity kit," which is filled with items such as a magnifying glass, a compass, and binoculars.

What Laws describes is a multidimensional approach to science—one in which students engage with scientific ideas through multiple representations including data, patterns, maps, words, and diagrams. As students engage in these different representations, neural pathways will be built that will allow different regions to communicate with each other, creating the kind of brain communication that was noted in the brains of "trailblazing people."

When I work with teachers, I often receive a very warm response to my call for the need to make teaching more multidimensional. This is then immediately followed by the pressing question: How do we teach this way when we have textbooks that we have to work through? Many teachers work in districts with prescribed textbooks whose authors are unaware of the value of multidimensional learning.

When teachers ask me this question, I suggest that they prune a page of repetitive questions down to three or four of the best questions and then invite students to engage with them differently, in the ways I suggest above. Any teacher can do this, it does not require new resources, and students can be invited to engage with the content in multiple ways in any subject area and at any grade level. When teachers start to work in these ways, they often become inspired and begin to think more creatively about their own subjects and the ways they teach them. This in turn creates more joy and fulfillment, especially as teachers experience the increased engagement of students.

My two daughters both attended a local public elementary school in Palo Alto. The teachers did not give a lot of homework, which I appreciated, as I know that homework has limited, if any, benefits and is often harmful to students' well-being.[14] When my daughters did receive math homework, it was often puzzles or KenKens (Japanese math puzzles), but occasionally they would bring home a worksheet of nearly identical questions. Often when they had worksheets of questions to complete, there would be tears and frustration. It continually baffles me why teachers think students should be working on repetitive and boring content in the evenings when they are tired. I do not work when I am tired, but my children are forced to when they are given homework.

I always try to be very supportive of my daughters' teachers, because I know teaching is one of the most demanding jobs possible and teachers are almost always wonderful and caring people. One evening, however, I decided I needed to intervene. My youngest daughter, around nine at the time, had come home with a worksheet of forty questions. She sat in front of the worksheet looking deflated. I was immediately concerned that this sort of worksheet would turn my daughter against math, so I asked her to work on only the first five questions. Then I wrote a note to the teacher on the worksheet that said:

> *I have asked my daughter to only work on the first five questions, and I can see that she understands them. I have asked her not to complete this worksheet, because I do not want her to think this is what math is.*

Other teachers laugh when I recount the message I sent to my daughter's teacher, probably feeling relieved that they are not my daughter's teacher. The positive news is that this particular event had a really good outcome. The fourth-grade teacher and I chatted about brain science and multiple approaches. Now, instead of worksheets, she gives four questions and asks students to solve them in different ways—she asks for a numerical solution, a story about the problem, and a visual solution. For my daughter, this was a great improvement over the boring, repetitive worksheets. There were no longer tears at homework time, and my daughter was happy to write her own story and draw her own picture. As she did so, multiple brain areas were involved and communicating with each other, and she was getting an opportunity to understand deeply.

Working in these multiple ways encourages brain communication while also bringing the content to life. The vast majority of students think about math as a set of numbers and methods and about English as books and words. When we approach math, English, science, or other subjects as opportunities for creativity and seeing things in multiple ways, it changes everything, stimulating vital brain growth and neural connections. Additionally, as teachers diversify the curriculum, moving from a simple list of numerical answers, pages of text, or scientific equations to visuals, models, words, videos, music, data, and drawings, the classroom changes from a place where all the work looks the same to one where the variety is enticing and creativity can be celebrated.

I like to illustrate the many different ways we can approach content by showing people a picture of seven dots. I tell them I'm only going to show the picture for a short time, and I want them to tell me how many dots there are. I ask them not to count them one by one, but rather to work out how many there are by grouping the dots; then I inquire how they grouped them. These are the seven dots.

Recently I asked a room of middle-school girls this question, and they found twenty-four different ways of seeing them! They wanted to keep going, but we were approaching lunchtime and I had to conclude the session. These were their twenty-four ways of seeing the dots.

I ask students to look at groups of dots in this way partly to illustrate the creativity in mathematics and the many different ways people see mathematics—even seven dots! I also do the exercise with students because it develops an important part of the brain called the approximate number system, or ANS. This is an area of the brain that allows a person to estimate nonverbally the number of a group of items. Students' levels of ANS proficiency have been found to accurately predict their future math achievement.[15]

A creative and multidimensional approach—the invitation to see in different ways—can be enacted in any content area. We could show students a scene from *To Kill a Mockingbird*, a cell diagram in biology, or an event from the news or from history and ask them: What do you see? How do you make sense of it? This values visual thinking and the multiple different ideas students will come up with, both of which are to be celebrated and encouraged at all times.

When Teachers Learn—and Use— a Multidimensional Approach

The Central Valley is a less well known area of California than the urban districts in the north (such as San Francisco) and the south (Los Angeles). When I first drove from Stanford to Tulare, about two hundred miles from Stanford and more than a hundred miles inland from the coast, I knew I was in the Central Valley when the scenery changed from roads with houses and shops to miles upon miles of cornfields.

The Central Valley is an agricultural region and an area of high need and low achievement. The educational leaders in Tulare County feel that the region is neglected in the opportunities teachers receive for professional development and for funding. Just over a year ago one of the county instructional coaches for mathematics, Shelah Feldstein, came to visit me at Stanford. She asked me about an idea she had to enroll all of the fifth-grade teachers in several districts in my online class "How to Learn Math." She also had wonderful plans to organize the teachers to take the course in groups and to sit together and process the ideas in school groups, with funded time.

Many amazing things happened over the next year that have been detailed in research papers,[16] but one I was particularly pleased with was that the teachers changed their own relationships with math. Fifth grade had been identified as the grade level with some of the lowest achievement; less than 8 percent of students met "proficiency" levels. In interviews at the end of the year, teachers admitted that they used to fear math time in their classes and try to get through it as fast as they could. After they learned about mindset, brain growth, and approaching problems multidimensionally, they enjoyed it so much, they would stay until seven p.m. in the evenings, discussing together how to visually approach problems.

Jim, one of the fifth-grade teachers, in an interview at

the end of the year described using one of our paper-folding activities and his surprise and pleasure when it allowed the students to think more deeply about exponents:

> They started folding the paper with triangles from a square. And on their own they discovered that there was an exponential relationship, so when they folded it once they had two pieces. And then when they folded it twice, they had four. And they started seeing the exponents of 2 with every fold. And they made that connection all by themselves, 'cause we've been doing base 10, and powers of 10. So I see those connections happening during these lessons, and that's huge for me.

The incredible change that the fifth-grade teachers went through, where they started seeing mathematics problems as opportunities to use multiple creative methods, was made possible because they also learned the brain science I presented in earlier chapters. Prior to taking the online class many of the teachers had a fixed mindset and did not feel that they were capable of coming up with or entitled to different ideas. When they were set free from this inaccurate and damaging thinking, they were able to approach mathematics and other subjects differently. One of the teachers reflected on her surprise that the online course had changed her as a person:

> I thought it was going to be great for the kids. I never expected it to change me. That's been my greatest revelation in all of it.

Not only did the teachers' own lives change; when they passed the ideas along to students, the students' lives changed as well. These changes were reflected in many ways, such as the students' changed beliefs about their potential and their learning. The students began to see their learning of mathematics as a different kind of activity. One of the teachers reported:

> The kids were thrilled, going, "Oh my gosh, he's doing it like that? It's okay that we struggle? It's okay we think differently?"

When students ask, "Is it okay that we think differently?" or "Is it okay that we struggle?" it shows me the damaging ideas they were harboring that had been holding back their learning. The idea that it might not be okay to struggle or to think differently is tragic, and yet these are ideas believed by millions of students, particularly about math. When the children changed their ideas about the value of struggle and learned to see math differently, the increase in their self-confidence was noticeable to the teachers, as Miguel reflected in an interview:

> I just want you to know this [the online course] has meant a lot. Seeing how positive the kids are about their learning now has made a world of difference. The confidence they have is unlike anything I have ever seen.

The students who changed their mindsets and their approach to learning, accepting struggle and being willing to see mathematics in different ways, received important benefits. Despite the narrowness of the state mathematics tests in place, at the end of the school year the students of teachers who took the course scored at significantly higher levels on tests of mathematics than students in other classes. The students who particularly benefited from the teaching changes and significantly increased their test-score performance were girls, language learners, and students from socioeconomically disadvantaged homes,[17] the students who often underachieve in mathematics and other subjects.

Jean Maddox was one of the teachers who was moved by the new knowledge she learned in the online class. She spent the year promoting the idea to her students that they can always grow and learn anything, and they should reject ideas suggesting their potential was fixed. For Jean the change to using visual methods was very important; it changed her own approach to mathematics as well as the ways she taught it:

> When I first started this journey, I was always doing the algorithm because that was my safety net. Now I'm thinking, "Okay, how am I gonna draw this? How do I visually see this?" Now I understand why the algorithm works, because I now have this totally clear picture in my head. Which has been a really good thing when it comes to things like fractions. And for these kids, it's like, "Oh, that's why it works." And for them to see that it was all this visual thing, and then somebody will go, "Oh my gosh!" So for some of those kids for whom math has been always having to memorize facts and that sort of thing, now it's like, "Oh!"

The changes the teachers went through illustrate the dual nature of being limitless—it is about changing your mindset and your ideas about yourself, but it is also about approaching subject content and life in a multidimensional manner.

One suggestion I made to the fifth-grade teachers was to drop questions with automatic answers and instead ask questions with answers that students are invited to find a way to reach. One of the teachers said:

The other day on the board I wrote: "The answer is 17. How many different ways can you get me to the answer?" I thought they might just say 1 plus 16, but they were doing orders of operations, they got really fancy, and I was really impressed with them.

A teacher mentioned on Twitter that she had also used this idea in her high-school geometry class. She put an answer on the board and asked students to use the geometric approaches they had learned to come up with the answer. She wrote that she was blown away by the different creative approaches the students arrived at, which were so valuable to the conversations that followed and the opportunities for brain connections.

Another of the fifth-grade teachers said that she now simply shows visuals of mathematical ideas and asks: "Okay, what do you see? What don't you see? What might you see? What could be the next thing?"

These different ideas are not complex, but they are at their heart all about multiple paths to learning. They encourage students' thinking to strike out in directions far different from "the usual." Teachers who make these changes in their methods are playing with content and experiencing the freedom that comes with this approach—instead of following textbooks, they experiment with ideas and invite students to experiment with them. We know now that teaching through a multidimensional approach also increases a brain's connectedness, which will help students develop into powerful and possibly "trailblazing" adults.

Other teachers have had experiences similar to those of the fifth-grade teachers in the Central Valley. Holly Compton still remembers being scared by one of her first experiences with math when she was told to work through a textbook page of multidigit arithmetic problems in first grade. She decided, as did her mother, that she did not have a "math brain." What followed for Holly were years of frustration and remediation. Holly's negative relationship with mathematics all started with a one-dimensional tool—a worksheet—that prompted her to decide that she was no good at math.

Math, sadly, has the potential to crush students' confidence more than any other subject. This is partly because of incorrect ideas about the ways it should be taught and learned, which produce first-grade experiences like Holly's. It is also because society says that people who can do math are "really smart" and those who struggle with it are "not smart." This can be devastating for many people, and Holly was one of these people. Unfortunately, I don't think the devastation Holly experienced is uncommon. Holly described the way her negative experiences in math impacted her entire life:

> It was really all-encompassing. My entire life was impacted by this lack of belief in myself.

Fortunately, Holly learned new ideas about herself and her ability to learn that caused her to become unlocked. A "game-changing" part of the process for Holly was seeing that math problems could be solved in different ways—showing the very important part that multidimensionality plays in becoming limitless. As Holly mentioned in an interview:

Now I view math as the most creative subject, because you can take things apart and put them back together, and you could have an hour-long conversation about 13 plus 12!

Holly retaught herself math and was encouraged by her students' thinking. As they started to approach math in different ways, she realized the subject was a different one from what she had thought. She started to experiment and be playful with math and saw her school's math scores improve. After some years of noteworthy teaching, Holly was invited to become the district's math coach—quite an achievement for a person who used to be terrified of math. Now Holly says that she is always teaching for a growth mindset, giving her students difficult multidimensional tasks, and telling them she hopes they can all struggle.

As well as causing her to teach differently, Holly's process of becoming unlocked has changed the ways she interacts with people, showing the additional benefits of a limitless approach to life. Holly used to go into meetings worried that she would not know something she needed to know—and feeling that she needed to be an expert. After becoming unlocked, she became less afraid of participating in meetings and more willing to take risks:

I am not afraid to say it. I will say to another teacher, "Hey, I'm stuck on this. Can you figure this out with me?"

This new openness to challenges and uncertainty seems to be a common reaction to becoming unlocked—people realize that it is good to struggle, that it is not a sign of a brain weakness, but of brain growth. This leads to more confidence in times of struggle and a willingness to share ideas that they are unsure of. One of the saddest, most central characteristics of fixed-brain thinking is the fear of being wrong. People's minds are literally locked, immobilized, by their fear, which is why an approach to life that values multidimensionality, growth, and struggle is so liberating. Holly said: "I have so many more ideas because I let myself have ideas."

Another core benefit of working and living with a multidimensional approach is that when roadblocks appear, you know there are alternate routes. Many of the adults I interviewed for this book said they would no longer stop when they met challenges or roadblocks; they simply would find another strategy, another approach. A multidimensional approach to knowledge reveals that there isn't only one way to do anything; there are always multiple ways forward.

The fact that Holly now feels freer to "have ideas" is so important. This is the sort of profound personal change that can come about when we understand the keys of learning. Whether we are working or studying in education or any other area, understanding the limits of traditional fixed-brain thinking and being empowered by our own ability to learn and grow is life changing in many different ways. This mind shift leads to greater self-confidence, resilience, and satisfaction in work and social relationships.

Holly shared that her relationships are now better, she has stopped doubting herself, and she is no longer depressed. Amazingly, this all came about when she was able to see mathematics and her relationship with mathematics differently.

For Holly a critical part of becoming limitless involved seeing math as a subject capable of being seen in multiple ways and valuing different people's ideas and ways of seeing. When you open your mind to see yourself and others as having endless potential, the impact is amplified when you also open content to different approaches. Multidimensionality is the perfect complement to a growth-mindset approach. Each one works better with the other in place.

Part of the huge success of our math camp, which resulted in an increase in student achievement by the equivalent of 2.8 years of school, came about because we used a multidimensional approach. When we interviewed students a year later, some of the students told us that they went back to classrooms where they were asked to do worksheets, but that they took the questions home and thought visually about them with their parents. One girl told me with regret that math class was not interesting now, because she was told she always had to follow the "teacher's method" and was not able to use her own methods. I was sad to hear this, but I also realized that the student now knew that there were multiple ways to think, not just the teacher's, and even though she was not able to use her methods, she was aware that they were important. She was frustrated, but her unlocked perspective was still operative.

In many classrooms students are given problems they do not know how to access—which causes them to think negatively about themselves and their learning. When problems are changed to become "low floor and high ceiling"—problems that are accessible by all but lead to more challenging work—everybody can access them and take them to different places.

We used these kinds of tasks in our camp, and we also valued multiple different ways of working, different ways of seeing problems, and different strategies and methods. We also encouraged rich discussions in which students shared their different ways of seeing and solving problems, and we all talked about them and compared the different approaches. For all of these reasons students were able to work productively and learn, and they could tell they were learning, so they were motivated to continue. We gave students clear access to the problems, and we gave them multiple examples of ways to explore answers. It is this combined approach of mindset and multidimensionality that is often missing in classrooms, homes, and offices.

It is similarly hard for students to become unlocked and develop limitless minds when they are in schools that give frequent tests and grades, as these also send fixed-brain messages to students.[18] The teachers I have interviewed for this book are different from many, because they understand the importance of students developing limitless minds, and to achieve this they combine brain and mindset messages with a teaching and assessment approach that enables growth and learning.

I use a multidimensional approach to teach math to my undergraduate class; we spend ten weeks together seeing mathematical ideas visually and sometimes physically as well as numerically and algorithmically—all of which create powerful brain connections. Here's an anonymous evaluation from one of my students:

> Math used to stay on the page, at least for me. Since I've started this class, problems have found their way into three-dimensional space. The walls of my room, the back of the name tag you asked me to make, my notebook for non-STEM classes—squares, diagrams, excitement converge in the brain space I had reserved for calculations; it used to be

one-dimensional, one-solution. The dimension I reserved for
mathematics has now explosively expanded.

Other students wrote in different ways about the ways that seeing mathematics visually and creatively and learning about brain science and mindset had given them a resilience that was changing their lives and giving them greater success in their other classes at Stanford.

Marc Petrie exemplifies a person whose life has changed dramatically as a result of learning about the benefits of struggle and of approaching content differently. Marc is in his sixties now, but when he was a young boy, he had an accident that left him partially disabled. His mother refused to accept that he would not recover and would have to attend special schools for the rest of his life. She took it upon herself to heal him by throwing bean bags to him to catch and developing his coordination. When Marc was a little older, learning to skate was a constant process of falling, getting up again, falling, and getting up again. He said that those early years of struggle gave him a growth mindset, because he would have "gotten nowhere without one." When Marc read about struggle in my previous writing, he immediately saw a connection to his own life and the ways that struggle had made him the person he is now.

Marc developed a growth mindset early in life, but it was a youcubed workshop that he attended a few years ago that gave him the language to talk about it with his eighth-grade students. Before Marc came to the workshop, he mainly taught from an uninspiring textbook. He returned to his classroom in Santa Ana after the summer workshop and changed his method.

He now starts class every Monday morning with a video of someone who has a growth mindset. On the day of our interview Marc had shown a video of a fifteen-year-old student who had developed a test for pancreatic cancer. His videos, which he finds from various sites on the internet, exemplify mindset thinking in action. Every Wednesday Marc presents his "favorite no"—a mathematical problem with mistakes that the students have to work to find. On Fridays students work on mathematics and art projects. As well as these regular class plans Marc takes a multidimensional approach to all of his teaching, encouraging students to make comics that illustrate mathematical ideas or showing them visuals of patterns or objects and asking them what they see. He told me that in both mathematics and art lessons, teachers project images and paintings and ask students what they see. He also invites students to build quilts of patterns and to explore the work of famous artists, looking at the symmetry in paintings, for example.

Before Marc made these changes, 6 percent of his students were meeting district levels of proficiency in mathematics. When he moved to the multidimensional, mindset approach, the success rate went up to 70 percent. Because Marc spoke of using so many wonderful and different ways to teach mathematics, through art, film, and other creative outlets, I asked him if he also worked from the textbook. Marc explained that he can achieve more when the students spend only twenty-five to thirty minutes—no longer—on "textbook stuff" and the remainder of the fifty-five-minute period on other projects. This makes complete sense to me.

Marc takes a growth-mindset approach not only to his mathematics teaching, but also to life. He explained that a few years ago when his son was young, his wife developed cancer and went through five different surgeries. Despite multiple surgeries and eighteen months of chemotherapy sessions, she continued to practice law. Marc had to be extremely strong during this time, as he had to take care of his wife and his son and teach. Marc said that he had to be "the most positive person I could be." Now his son is in college and his wife is in recovery, and Marc and his wife spend their Saturdays baking cookies for those in shelters. His wife also knits hats for women who are in chemotherapy. Marc relayed a way of thinking that was similar to that of others I interviewed who had become limitless. This is the approach of turning something negative into something positive. He talked about a concept in Judaism called *tikkun olam*, "healing the world," and how he sees this as related to having a growth mindset. Marc reflected: "It's almost, to me, 'Why am I on this planet? Why am I here? Why am I in this classroom?' There has to be a reason."

Marc's positive approach to life, even in times of extreme adversity, is inspiring. The changes he made in his classroom have resulted in huge gains in student achievement. It has affected other teachers in his school as well. After the sixth- and seventh-grade teachers saw Marc's success in eighth grade, they started to follow some of his ideas, and they too saw significant increases in their own students' achievement levels.

Anyone can learn content from any subject with a multidimensional approach. Appendix I, at the end of the book, offers readers an opportunity to think visually about mathematics. Learners who are in classrooms that do not approach content multidimensionally can take this approach on their own. I have already talked about the summer camp we conducted with eighty-three middle-school students who came to the Stanford campus. When we followed up with the students a year later, one of the boys told us that he understands volume more deeply now because he always thinks back to what a one-centimeter cube looks and feels like from an activity we worked on with sugar cubes. It is unfortunate that those students did not get to continue with opportunities to think visually, physically, and in multiple other ways at school, but the eighteen days had, fortunately, given them a different perspective on learning that they were able to take into their lives.

Leah Haworth, one of the teachers I interviewed, talked about the enormous changes that came about for her students when, instead of ruled exercise books, she gave them expansive blank journals and told them to use the journals to play with ideas, to draw ideas, and to think. Giving students a creative space in which to think and explore is perfectly aligned with a multidimensional approach to content.

A few years ago, I was conducting a trial of our week of inspirational math tasks in a local school. These are a set of visual and creative K–12 mathematics lessons that we provide free on our youcubed site for anyone to use. I was walking down the corridor after one of the lessons when a mother of one of the girls rushed up to me. She asked me what we had been doing in her daughter's math class for the last few days—her daughter, who had always hated math and could never learn it, had changed her mind! She now saw a future for herself in math. This was really good to hear, as I know that when children change their minds about what is possible and open their hearts to a different approach, their learning pathways change.

The first three learning keys, which address the value of understanding growth and challenge, are critical to unlocking learning potential. However, others can find these messages frustrating and counterproductive without the proper context for creative brain development. When a growth mindset meets the restrictions of the fixed-brain world, it loses some of its potential for change. The answer, we now know, is multidimensional learning—Learning Key #4. Taking a multidimensional view of a problem, a topic, or the world in general unlocks our ability to learn and grow in vital ways. A growth mindset together with multidimensional learning opportunities will allow learners of any age to break free of fear and to overcome obstacles, to see problems with a fresh perspective, and gain confidence in their own ability. Even when we are working inside rigid, fixed systems that do not value the multiple ways people think—whether it is test-driven schools or workplaces that value only narrow perspectives—taking a multidimensional approach to problems we face will support and enforce all aspects of learning and living.

5

WHY SPEED IS OUT
AND FLEXIBILITY IS IN!

INCORRECT IDEAS, faulty methods, and false assumptions restrict learning potential in so many ways. The good news is that we now have the science and a wealth of proven contrasting approaches that unlock learning and potential. We have discussed two of the major damaging myths—the idea that brains are fixed and that struggle is a sign of weakness. When people let go of these faulty ideas, they change in profound and generative ways.

This chapter deals with another harmful myth and offers its liberating counterpart—the idea that to be good at math or any other subject, you need to be a fast thinker. When we let go of the idea that speed is important and approach learning as a space for deep and flexible thinking, it enables a breakthrough in the ways we encounter the world. Creative and flexible thinking is the kind employed by "trailblazers" in their fields,[1] as I mentioned in the last chapter, and is available to us when we approach knowledge with a new perspective.

LEARNING KEY #5

Speed of thinking is not a measure of aptitude.
Learning is optimized when we approach ideas,
and life, with creativity and flexibility.

Mathematics, more than other subjects, has been damaged by the idea that to be good at the subject you need to be fast. This has come about in part because of damaging school practices, such as timed tests of math facts, often given to children as young as five years of age. Parents also use speed-based math activities such as flashcards with their children. These are all part of the reason that most people associate math with speed, thinking that if they are not speedy with numbers, they cannot be successful. I show images such as this worksheet to audiences.

Multiplying by 12   NAME _____

2 x12	12 x12	6 x12	7 x12	6 x12	12 x12	4 x12	8 x12	2 x12	5 x12	12 x12	4 x12
9 x12	4 x12	12 x12	2 x12	3 x12	3 x12	6 x12	4 x12	11 x12	6 x12	7 x12	2 x12
1 x12	8 x12	5 x12	12 x12	9 x12	7 x12	11 x12	6 x12	2 x12	2 x12	7 x12	12 x12
7 x12	5 x12	1 x12	12 x12	8 x12	6 x12	8 x12	3 x12	0 x12	6 x12	4 x12	2 x12
5 x12	12 x12	4 x12	2 x12	6 x12	11 x12	4 x12	9 x12	3 x12	8 x12	3 x12	2 x12
6 x12	4 x12	12 x12	12 x12	12 x12	0 x12	9 x12	4 x12	8 x12	5 x12	2 x12	7 x12
5 x12	1 x12	8 x12	12 x12	7 x12	4 x12	12 x12	5 x12	9 x12	1 x12	3 x12	7 x12
8 x12	9 x12	5 x12	5 x12	6 x12	11 x12	7 x12	3 x12	6 x12	5 x12	8 x12	5 x12

Goal _____   Number Correct _____

They are often met with groans, although a few people (a small minority) say they enjoyed taking the tests. We now know that timed math-fact tests given to young children are the beginning of math anxiety for many, and new brain research helps us understand the process by which this happens.

The Effects of Stress and Anxiety

Neuroscientist Sian Beilock has studied the brain when people are working under pressure. A particular area of the brain called the "working memory" is needed when we do calculations. The working memory is sometimes referred to as the "search engine of the mind" and, like all areas of our brains, is developed through practice. What Beilock has shown is that when we are stressed or under pressure, our working memory is impeded.[2] The students who are the most compromised are those with the most working memory. This means that when students are given timed math tests and they become anxious, as many do, their working memory is compromised, and they cannot calculate the answers. Anxiety sets in, and a pattern of harmful beliefs soon follows.

The feeling of stress impeding your brain may be something you have known yourself. Have you ever had to work on a math calculation under pressure and felt as though your mind "went blank"? That is the feeling of stress blocking your working memory. When we give timed tests to young children, many of them experience stress, their working memory is compromised, and they cannot recall math facts. When they realize they cannot achieve, anxiety sets in.

I have taught Stanford undergraduates for many years, and every year a substantial proportion of my students have this anxiety and fear. I always ask those who are mathematically traumatized what happened and when. Almost every student I have asked has responded in the same way—recalling the timed math tests they took in second or third grade. Some became anxious and did not do well while others did well on them, but the tests caused them to think (not surprisingly) that mathematics was a subject of shallow recall, and they turned away from math.

Teacher Jodi Campinelli described a set of devastating events she went through as a young child who struggled with timed tests. Jodi was told at the end of second grade that she might have to repeat the year, because she did not do well on timed tests. This first part of the story horrifies me, but there is more. She was told she would have to go to tutoring with the principal, which Jodi describes as "torture." To add to this, her parents made her do timed tests in the kitchen in the evenings; they set up a timer next to her that ticked away as she worked feverishly on the calculations.

I hate to think of this second-grade child undergoing this kind of stress. She was given the idea that math-fact tests were an indication of her intelligence, her worth as a person, and then told she was failing. Jodi often did not finish the tests in the time her parents gave her, or if she did, she made mistakes, after which her mother told her it was okay, as she wasn't good at math either. Jodi said that the sound of a kitchen timer still "freaks her out" to this day, and I am not surprised to hear that.[3]

Jodi received many negative messages when she was in second grade, and her mother's reassurance that she was also bad at math, given no doubt with the best of intentions, was one of them. Sian Beilock's research has revealed interesting associations that tell us just how damaging such messages are. In one study, she and colleagues found that the amount of math anxiety expressed by parents predicted their child's achievement in school.[4] The amount of math knowledge parents had was not important, only how anxious they were. And their math anxiety impacted students negatively only if parents helped with homework. Apparently if parents are anxious about math but never interact with their children in actually doing math, the anxiety is not passed on. If they are helping with homework, they are probably giving the message that math is hard or they were not good at math or, worse, setting their children up with timers in the kitchen.

Beilock and her team also found that the amount of math anxiety female elementary-school teachers have predicts their female—but not male—students' achievement.[5] I imagine this result comes about because female teachers share their feelings about math through statements I have heard like the obvious, "I was not good at math in school," but also, "Let's just get through this quickly, so that we can move to reading time." Girls are affected by this more than boys, because they identify more with their same-gender teachers. Both studies show that parent and teacher messages about math can reduce students' achievement. This highlights again the relationship between our beliefs and our achievement.

Fortunately, Jodi had better experiences as she moved forward in school. She learned about the damage of emphasizing speed and is now a middle- and high-school math teacher sharing the message that slow, deep thinking is what is important. Over time she has learned that the timer in her kitchen was not measuring her worth, and she has gone through an important process of becoming limitless. The knowledge that speed is not important was a particularly important key for her.

The Neuroscience of Speed

The irony of the unfortunate speed-based math activities in schools, where children are turned away from a lifetime of mathematical and scientific thinking because they don't produce math facts quickly and under pressure, is that mathematics is not a subject that requires speed. Some of the strongest mathematical thinkers are very slow with numbers and other aspects of mathematics. They do not think quickly; they think slowly and deeply.

In recent years, some of the world's greatest mathematicians, including those who have won the Fields Medal, such as Laurent Schwartz[6] and Maryam Mirzakhani,[7] have talked openly about how slow they are with math. After Schwartz won the Fields Medal, he wrote an autobiography about his school days in which he talked about feeling stupid in school because he was one of the slowest thinkers. He says:

> I was always deeply uncertain about my own intellectual capacity; I thought I was unintelligent. And it is true that I was, and still am, rather slow. I need time to seize things because I always need to understand them fully. Toward the end of the eleventh grade, I secretly thought of myself as stupid. I worried about this for a long time.

I'm still just as slow. . . . At the end of the eleventh grade, I took the measure of the situation and came to the conclusion that rapidity doesn't have a precise relation to intelligence. What is important is to deeply understand things and their relations to each other. This is where intelligence lies. The fact of being quick or slow isn't really relevant.[8]

When I was in school, I was a fast thinker, to the great annoyance of my tenth-grade math teacher. She would start class each day by writing about eighty questions on the board. I would amuse myself while she was doing this by working out all the answers as fast as she could write out the questions. By the time she put her chalk down and turned to face us, I had finished all of the questions and would hand her my paper. She never looked pleased, and one time told me that I was just doing it to irritate her (there is a lot to think about in that statement). She would look through all of my answers hoping I had made a mistake, but I don't remember ever making one. If I could go back in time with the knowledge I have now, I might point out to my teacher that I was finishing the questions so quickly because they did not require any deep or complex thinking, although that would probably not have gone down well.

At the time I was speeding through my math questions, I was myself working under the myth that speed is what is important. In our archaic school system, it is not surprising that millions of students believe speedy performance is what is valued. Now many years on, I have learned to approach content differently. I no longer look at math problems as something to answer quickly, but as something to think about deeply and creatively. That change has helped me greatly. I now get more from not only mathematical thinking, but any scientific or technical reading or work. The change in my approach has helped me so much and has fueled my passion to help others disarm this pervasive myth in the pursuit of understanding, creativity, and connections.

Medical doctor Norman Doidge says that when people learn something quickly, they are probably strengthening existing neural connections. These he describes as "easy come, easy go" neural connections, which can be rapidly reversed.[9] This is what is happening when we study for a test, and we go over something we have already learned. We cram information in and reproduce it in a day or so, but it does not last and is quickly forgotten. More permanent brain changes come from the formation of new structures in the brain—the sprouting of neural connections and synapses. This is always a slow process.

Doidge refers to a study of people learning Braille. Researchers saw the quicker brain development start immediately, but the slower, deeper, more permanent development took much longer. It was this learning that lasted and that was still in place months later. Doidge counsels that if, as a learner, you feel that your mind is like a sieve and you are not learning, keep going, as the deeper, more effective learning will come. He says that the "tortoises" who seem slow to pick up a skill may nevertheless learn it better than their "hare" friends, the "quick studies" who won't necessarily hold on to what they have learned without the sustained practice that solidifies their learning.[10]

When people learn slowly and quickly, teachers often assume they have different potential, but they are actually involved in different brain activity, and the slow, deep activity is more important. Schools in the US tend to value the faster, more shallow learning that can be assessed through tests, and students who quickly memorize are usually success-ful using these measures. And yet the research shows that the students who struggle more and learn more slowly are achieving the most in the long term.

One of the ways that fast learning damages us is when slower learners compare themselves to those working more quickly. This usually results in feeling inadequate at the task at hand. In schools and colleges across the country, students regularly write themselves off when they see others working more quickly. Nancy Qushair, who is the head of the math-ematics department at an IB (International Baccalaureate) school, described a common occurrence: a student who had given up on herself because she saw others "getting math" more quickly. When Millie started in Nancy's class, she said she hated math and described herself as "dumb." She wrote a note to Nancy with these words:

> I would look at kids sitting next to me and they'd finish so much faster than me. And they'd already be done and I could barely get started. And I always compared myself to them and I always thought, "I'm never gonna be able to do this."

Millie is not alone in these feelings, and they are feelings that we now know cause our brains to function less well. Nancy decided to change Millie's trajectory, which she did very purposefully and carefully. She asked Millie to only fo-cus on herself, not others, and to set herself a goal—one thing she would like to achieve in the next few weeks.

Millie said she "finally" wanted to understand integers.

"Okay," Nancy said. "We're not here to know all the material for the rest of the year. We're just gonna know integers. So we're gonna work together."

Nancy gave Millie a range of visual representations—number lines, thermometers, and pictures of Prada purses—to think about mathematically. And throughout the year she found ways for Millie and the other students to work more creatively. By the end of the school year, Millie was a changed person, and she ended the year by writing this note to Nancy:

> Dear Ms. Qushair, I just wanted to thank you so much for being a great teacher. I'm not just saying that you were a great person. I'm saying that you are truly a good teacher. At first, I thought your videos about people not being able to do math were wrong. I truly thought I couldn't do math. I didn't realize if I thought like that I would never be able to get any further. You as a teacher not only taught me the subject, but taught me how I see things and how I do math. I'm a creative person, so math was never my thing. When you started doing visuals and showing us why we do it instead of just how, I was starting to get it. I knew that once I got it, I had to keep going. You helped me so much with all of that. It's been almost a year now, and I feel like I've grown so much. I really didn't think I could make it this far. "Just try it, Millie." You would always say that and I thought, "I could try, but I won't, I won't succeed." I was very wrong. You knew that I could do it, and that helped me get through the whole year, so I just wanted to say thank you.

Millie says something very illuminating in this note. She talks about what we know to be so important—the fact that

Nancy believed in her and kept giving her positive messages. But she also says, "When you started doing visuals and showing us why we do it instead of just how, I was starting to get it. I knew that once I got it, I had to keep going." Here Millie captures a core component of learning that I talked about in the last chapter—it is not enough to share positive messages with students; we have to give them access to understanding and experiences of success.

This brings us back to the multidimensional approach, in which the learning process is opened up and creative and visual tasks are given to help students see mathematical ideas differently and be successful. This approach is far more effective than the shallow, rote memorization techniques used in the past. And yet, in many areas, we continue to promote memory skills, even though we now know that good memorizers do not have more math potential.[11] The memorizers find they can be successful by following teachers' methods, often without developing any understanding. I have met many successful mathematics students, even those majoring in math at top universities, who lament that they have really understood none of the work they have completed over the years. When we value memorization over depth of understanding, we harm the deep thinkers who turn away from the subject. We also harm the successful memorizers who would have been helped by an approach to knowledge that gave them access to deep understanding.

When Nancy made mathematics visual and gave students insights into "why we do it instead of just how," Millie experienced her first success. And once she experienced that, she kept going, and she started to reject the myth she had believed—that she could never be good at math.

Nancy not only worked to create a positive mathematics experience for all of her class and other classes in the school, but she saw students who had given up on themselves and worked to change their experiences. She gave Millie work to do at home that she knew would help her see math differently. She even arranged during one of the tests at the school to sit with Millie and show her how she could answer the questions with visual thinking. Before attending Nancy's class, Millie received Ds and Fs in math. At the end of the year with Nancy, she got a B. More important, she understood the material and no longer believed she was bad at math.

Nancy has now worked with all of the teachers in the school to learn the ideas that I share in this book, in my *Mathematical Mindsets* book, and in my online classes. Reflecting on the changes in the whole school, she said:

> I never thought I would come to that day to finally see a group of teachers passionate about teaching the kids and passionate about what they're teaching in math. And I can't wait to see the change in their students. It's not just my class. It's not just one student. There are so many kids. There are so many teachers who have really benefited, and it's impacted them in their daily lives.

The wonderful story of the transition Nancy worked to bring about in Millie, a shift in perspective and approach that will change Millie's life, highlights some of the important ways learners can develop limitless minds. To better understand this concept, we are going to take a deeper dive into the world of math learning with the results of a fascinating study that has implications for all teachers and learners, of any subject, as well as parents and leaders. The study gave interesting insights into the working of the human mind and the role of flexibility.

Thinking Flexibly

Two British professors, Eddie Gray and David Tall, both at the University of Warwick, worked with groups of students ages seven through thirteen who had been categorized by their teachers as low-, average-, or high-achievers.[12] The researchers asked the students numerical questions, showed them visuals, and collected their strategies. For example, they gave the problem 7 + 19, showing the numbers visually.

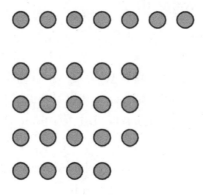

The researchers found that the difference between the high- and low-achieving students was not that high-achieving students knew more, but that they engaged with numbers flexibly. The researchers categorized the different strategies: in "counting all," students simply counted all the numbers; in "counting on," they started with a number and counted upward; "known facts" indicated when students remembered math facts; and "number sense" indicated when students used numbers flexibly, for example, approaching an addition problem such as 7 + 19 by working out 6 + 20 instead. These were the strategies the researchers collected from the high-achieving and low-achieving students:

High-achieving students:

30 percent known facts

9 percent count on

61 percent number sense

Low-achieving students:

6 percent known facts

72 percent count on

22 percent count all

0 percent number sense

The results were dramatic. The high-achieving students were thinking flexibly; 61 percent of them used number sense. This strategy was completely absent among the low-achieving students.

The low-achieving students had developed counting strategies, such as counting on or counting back (starting with a number and counting down), and they clung to these strategies, using them for all questions, even when they really did not make sense. The researchers made an important point: the low-achieving students, who were not thinking flexibly, were learning a different mathematics, and the mathematics they were learning was more difficult. They illustrate this with the problem 16 – 13.

The low-achieving students approached 16 – 13 by counting back, which is actually rather difficult to do (try starting at 16 and counting down 13 numbers), with a lot of room for error. The high-achieving students dealt with the numbers flexibly, subtracting 3 from 6 and 10 from 10 to get 3. This kind of number flexibility is extremely important, but when students are trained to memorize math facts blindly and work with algorithms before they understand them, they automatically resort to memorization and never develop the ability to think of numbers flexibly.

Students who appear to be low achieving in the early years of school, particularly with number problems, are often pulled aside and drilled—a "drill and practice" approach aptly renamed by many students as "drill and kill." This is probably the last thing that they need. They are underachieving because they have the wrong approach to mathematics, thinking they need to use memorized methods. They have memorized counting strategies that they keep using, even when number sense would be much more helpful. They need, instead of being drilled, to engage with numbers flexibly and creatively. They need to approach numbers differently.

Conceptual Learning

What does it mean to approach numbers conceptually? This may be a foreign idea for many readers who have always been encouraged to approach numbers as methods, facts, and rules. Gray and Tall distinguish between concepts and methods in early mathematics as shown in the following chart.

We learn methods such as counting in order to develop the concept of a number. We learn to "count on" in order to develop the concept of a sum, and we learn repeated addition

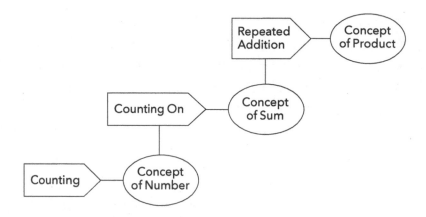

in order to develop the concept of a product. Mathematics is a conceptual subject, but many students do not learn it conceptually; they learn it as a set of rules or methods to memorize. As discussed, this turns out to be a serious problem for many students, and some fascinating research on the brain sheds light on the reasons why.

When we learn new knowledge, it takes up a large space in the brain—it literally occupies more room—as the brain works out what it means and where it connects with other ideas already learned. But as time goes on, the concepts we have learned are compressed into a smaller space. The ideas are still there so that when we need them, we can quickly and easily "pull" them from our brain and use them; they just take up less space. If I were to teach arithmetic to kindergarten students, the concepts would take up a large space in their brains. But if I asked adults to add 3 and 2, they would quickly do so, pulling the answer from their compressed knowledge of addition. William Thurston, a mathematician who won the Fields Medal, described compression in this way:

> *Mathematics is amazingly compressible: you may struggle*
> *a long time, step by step, to work through the same process*

or idea from several approaches. But once you really under-
stand it and have the mental perspective to see it as a whole,
there is often a tremendous mental compression. You can file
it away, recall it quickly and completely when you need it,
and use it as just one step in some other mental process. The
insight that goes with this compression is one of the real joys
of mathematics.[13]

You may be thinking that few students describe math as a "real joy," and part of the reason is that we can only compress concepts. So when students are engaging in mathematics conceptually—looking at ideas from different perspectives and using numbers flexibly—they are developing a conceptual understanding, creating concepts that can be compressed in the brain. When students believe that mathematics is about memorization, they are not developing a conceptual understanding or forming concepts that can then be compressed.[14] Instead of compressed concepts in the brain, their math knowledge is more like a ladder of memorized methods that stack one on top of another, stretching, as it may seem to these learners, to the sky.

When I tell teachers and parents about this research, they ask me: "How do I get my students to learn conceptually?" There are many ways to engage students conceptually. First, it is important to give students access to the reasons why methods work, not just give them methods to memorize. In the last chapter I spoke of the value of asking students how they see an idea, which can really help with understanding it conceptually.

Another conceptual approach to the teaching and learning of numbers, called "number talks," was devised by educators Ruth Parker and Kathy Richardson and developed by Cathy Humphreys and Sherry Parrish. The method involves talking about different approaches to number problems. In a number talk, students are asked to work out a number calculation in their heads, without using paper and pencil, and then teachers collect their different methods. When I teach others to use number talks, I also recommend they collect visual representations, to encourage different pathways in the brain to be activated. To understand this more deeply, try working out 18 × 5 in your head before reading or looking ahead to some solutions.

Here are six different ways of calculating 18 × 5 (there are more) with their visual representations.

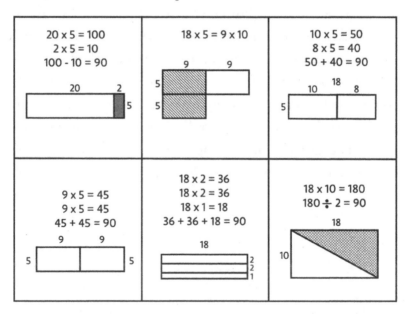

We can pose all sorts of different number problems and solve them in different ways, breaking numbers apart and making them into "friendlier" numbers, such as 20, 10, 5, or 100. This makes calculations easier and encourages number flexibility, something that is at the heart of number sense. We should celebrate with students the many different ways mathematics can be seen and solved, instead of teaching mathematics as lists of methods to remember.

When I have shown the different ways that people can approach even a bare number problem to various audiences, many people have expressed surprise and felt a sense of liberation. One day I was invited to meet with the amazing professor and inventor Sebastian Thrun and his team at Udacity. Thrun invented self-driving cars and was one of the first creators of massive open online courses (MOOCs). He is now working on designing flying cars. I interviewed him in my first online teacher course to help spread his ideas about mathematics and teaching.

When I first met Sebastian, he invited me to Udacity to chat with his team. I sat, that day, in a crowded room packed with his engineers. Those of us who fit sat around a big table; others stood against the walls. Sebastian asked me about good approaches to math, so I asked the room of people if they would like to do a math problem together. They eagerly agreed. I asked them to solve 18 × 5 and then I collected their different methods, noting them on the write-on table and showing the visuals as I went. The group was amazed, so much so that some of the team immediately went out onto the streets and started interviewing passersby, asking people to solve 18 × 5. They then made a mini online 18 × 5 course and made 18 × 5 T-shirts that they started wearing around Udacity.

I shared the same approach with another amazing tech leader, Luc Barthelet, who led the product development of *SimCity*, who at the time was the executive director of Wolfram Alpha, an online computational data site. He was so excited about it, he started asking everyone he met to tackle the problem. Of course, 18 × 5 is not the only problem that can be solved in multiple ways. These different people—high-achieving mathematics users—felt liberated when they saw that mathematics problems could be approached in multiple creative ways.

Why are people so surprised by this multidimensional, creative approach to mathematics? One person who had gone through the 18 × 5 exercise was shocked. She said, "It isn't that I didn't know you could do that with numbers, but I had somehow thought that it was 'not allowed.'"

A teacher from England related his experience with number talks. He tried a number talk in his "top set" (top track) group and started with the problem 18 × 5. The students had willingly shared different methods for solving 18 × 5 and engaged in a good discussion. He then asked his "bottom set" group the same question, but was greeted by silence. The students could work out the answer with an algorithm, but had no other approaches. He suggested that the students think of other ways, for example, working out 20 × 5. The class was shocked and said to him, "But sir, we thought we were not allowed to do that." The high-achieving students had learned to approach numbers with flexibility, and the low-achieving students had not; they thought that number flexibility was "not allowed."

This speaks to the damage that has been done in mathematics education—people think that number flexibility is not allowed and that math is all about following rules. It is no wonder so many people turn away from the subject. This is a problem I have noticed on multiple occasions. It is a problem for all students and for our nation, but it seems to be a particular problem for low-achieving students, as the teaching example and research from Gray and Tall suggest.

One particularly useful approach in mathematical problem solving is called "taking a smaller case." When we approach a complex problem by trying it with smaller numbers, the patterns inherent in the problem often become clearer and more visible. Consider, for example, the elegant proof that came to be known as Gauss's proof. This is one of those lovely mathematical patterns that can be seen in all sorts of situations and is helpful for everyone to know about, whether teaching or parenting children taking math or not.

Carl Friedrich Gauss was a German mathematician who lived in the 1800s. I don't know how accurate this often-told story about Gauss as a child is, but it is a good story! When young Gauss was in elementary school, his teacher realized she needed to give him challenging problems, so she gave him a problem that she thought would take him a long time. She asked him to add every number from 1 to 100. But young Gauss saw interesting patterns and realized he did not need to add all the numbers. He noticed that if he added 1 and 100 he would get 101, and if he added 2 and 99 he would get 101, and if he added 3 and 98 he would get 101. He saw that he had 50 pairs equaling 101 and the total was 50 × 101.

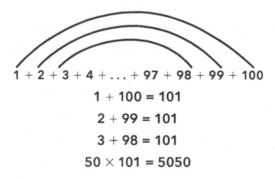

$$1 + 2 + 3 + 4 + \ldots + 97 + 98 + 99 + 100$$
$$1 + 100 = 101$$
$$2 + 99 = 101$$
$$3 + 98 = 101$$
$$50 \times 101 = 5050$$

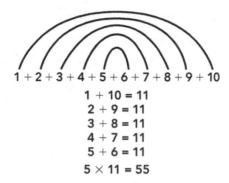

$$1 + 2 + 3 + 4 + 5 + 6 + 7 + 8 + 9 + 10$$
$$1 + 10 = 11$$
$$2 + 9 = 11$$
$$3 + 8 = 11$$
$$4 + 7 = 11$$
$$5 + 6 = 11$$
$$5 \times 11 = 55$$

To understand the patterns in Gauss's proof, it helps to take a smaller case—for example, looking at what happens with the numbers 1 to 10.

This smaller set of numbers helps us see what is going on and the reason that pairing numbers that are one bigger and one smaller than the previous pair will give us the same number. If you would like an extra challenge and an opportunity for brain growth, consider how Gauss's method can work with an odd number of successive numbers.

Taking a smaller case is an inherently mathematical act, but when I have taught it to students, low-achieving students resist. I understand why. They have been taught that mathematics is a set of rules to follow. The idea of not answering the question given to you but asking a different one, adapting the question, is completely foreign to them and seems to break the "rules" they have learned.

Being taught to play with numbers and see mathematics as a subject that can be treated in an open and multidimensional way seems to me to be a critical approach to life. I say this not to be overly dramatic, but because I know that when people see mathematics differently, they see their own potential differently, which changes their lives and offers access to experiences that they would not have had otherwise.

It not only allows them to be successful in mathematics and go forward with STEM subjects in school and beyond; it gives them a quantitative literacy that will help them understand finances, statistics, and other math-related areas of their lives.

When teachers give students an experience of mathematics that is open, conceptual, and untimed, it is incredibly freeing. The story of one student liberated by a conceptual mathematics problem comes from Nina Sudnick, a fourth-grade teacher in Ohio. During her first year of teaching, Nina was shocked by how little her students knew, despite the fact that it was their fifth year studying math. Nina sought to understand better and ended up reading one of my earlier books, *What's Math Got to Do with It?* She recalled:

> *I was reading that book, and if I show it to you, I think almost every sentence is underlined. My brain was exploding because of different ideas that had always bothered me, but I never could really articulate them. I couldn't understand why these students had had such difficulty.*

Nina returned to school after the summer and changed her teaching. At the end of her first year, 64 percent of her students had scored at proficiency levels. The next year, after Nina changed her teaching methods, that figure went up to 99 percent.

One of the important changes Nina made was in the way she handled her daily and weekly assessments. Instead of grading tests with pluses and minuses and giving them back to students, a process that conveys fixed messages to students about their achievement, Nina started to write comments on the students' work, pointing out what they understood and what they were only beginning to understand. At first when students got their papers back, they were searching for pluses and minuses, but they couldn't find any. Nina says that she now views their performance on tests as an indication of where they are on a spectrum of understanding.

Nina also gave students more open and conceptual mathematics problems. One of the problems she gave her students was from our "Week of Inspirational Math"—a set of open and creative problems we share each year on youcubed. This was an unsolved problem in the history of mathematics called the Collatz conjecture, and we posed it in this way:

- Start with any whole number.
- If the number is even, divide it by 2 (halve it).
- If the number is odd, multiply it by 3 and add 1.
- Continue generating numbers until your sequence ends.
- Choose another number and create the sequence. What do you think will happen?

No one has ever found a sequence of numbers that does not end in 1 or proved why this happens. The problem is also known as the hailstone sequence, because of the pattern the numbers make, behaving like hailstones in a cloud in the way they go up and then back down again.

Raindrops are pushed by the wind above the freezing point, where they spin around, freeze, and grow until they are heavy enough to fall to the earth as hailstones.

Even though nobody has ever solved this problem, we decided it was suitable for third graders and older. Many teachers gave it to students, challenging them to be the first person to find a pattern that did not end in 1, which of

How Hailstones Are Formed

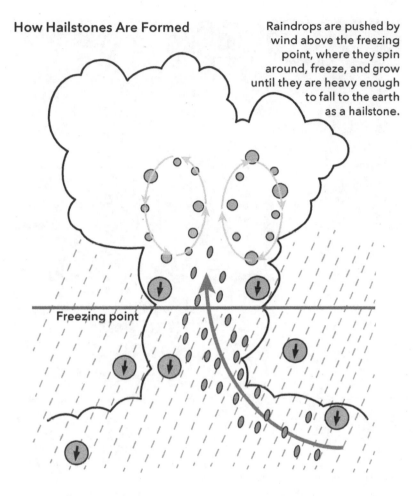

Raindrops are pushed by wind above the freezing point, where they spin around, freeze, and grow until they are heavy enough to fall to the earth as a hailstone.

Freezing point

course they loved. On Twitter we saw many photos teachers had posted of students producing beautiful representations of the patterns.

One of Nina's students was Jodi, a girl whose medical condition caused her to miss a lot of school during the year and often kept her from completing homework. Math had never been a subject she enjoyed, but she became enchanted by the hailstone problem. One day Nina noticed that Jodi was walking around with her pockets stuffed full of small pieces of paper. Over a few weeks, Nina saw the pockets grow-

Visual Representations of Number Patterns

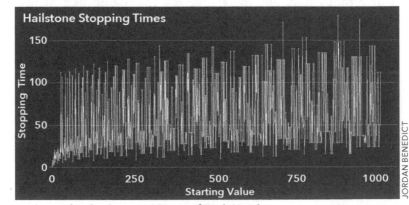

Students Plot the Stopping Times of Each Number

ing bigger and bigger till the pieces of paper began spilling out. Eventually Nina asked her what they were. Jodi reached into a pocket and handed Nina her scribblings of different patterns she had been trying. Nina had been working diligently for weeks on the Collatz conjecture, trying pattern after pattern. Nina reflected:

> *She knows the pattern, and she is so darn proud of that, Jo. And I'm like, "I don't care if you do any homework the rest of the year. [chuckle] Just keep working on your hailstone sequence." A lot of the kids were saying, "Gosh! Every time 16 comes up, it's the same pattern." I'm like, "Oh, really." Jodi felt successful. Probably the first time in all her years in mathematics, so thank you for that.*

The impact of our education systems' reliance on memorization rather than conceptual learning in mathematics was shown clearly in a recent analysis conducted by the Programme for International Student Assessment (PISA) team. The PISA, a test of problem solving run by the Organisation for Economic Co-operation and Development (OECD) in Paris, is given every three years to fifteen-year-olds across the world. I was asked to visit the PISA team in Paris a few years ago to help them with the analyses they were conducting.

I pulled a seat up to the table on my first morning in Paris, and the first thing the PISA team said to me was, "What's up with Americans and pi?" They were referring to the fact that on all of the questions involving pi (the ratio of a circle's circumference to its diameter, an irrational number that begins 3.14) American students did abysmally, coming in last or close to last in the world. I had an answer for them.

Having moved to the US from the UK, I had always noticed something curious about the teaching of pi. In the US, children are taught to memorize as many of the digits as possible. The number is often shortened to 3.14 but extends infinitely. This leads US students to think of pi as a "number that goes on forever," which seems to obscure its real meaning as a relationship between the circumference and diameter of a circle. The relationship it represents is actually dynamic and fascinating, because no matter what size circle we measure, the ratio of the circumference to the diameter is always the same.

I recently asked teachers to ask their students the meaning of pi, to see what they would say. Sure enough, the teachers reported to me that their students all said that pi was a really long number. No one mentioned circles or the relationships inside them. It's no wonder that students did badly on all the circle questions on the PISA. There is nothing wrong with having fun with pi and asking students to memorize digits (or to eat pie!), but such activities should be accompanied by deeper investigations of circles and their relationships.

That year, 2012, the PISA team conducted an analysis not only of students' test scores, but also of their approaches to learning. In addition to math questions, the team gave students survey questions, asking them how they learned. The approaches fell into three general categories. In the memorization approach, students tried to memorize everything they were shown "by heart." In the relational approach, students related new ideas to those they already knew. And in the self-monitoring approach, they evaluated what they knew and worked out what they needed to learn.

In every country the students taking a memorization approach were the lowest achievers, and countries that had high numbers of memorizers—the US was one of these—were among the lowest achieving in the world.[15] In France and Japan, for example, pupils who combined self-monitoring and relational strategies outscored students using memorization by more than a year's worth of schooling. The study showed, on an international level, that taking a memorization approach to learning does not lead to high achievement, whereas thinking about ideas and relationships does.

As we have seen from the research, mathematics teaching in the US is badly broken. Mathematics can be a beautiful subject of ideas and connections that can be encountered conceptually and creatively, but students who attend schools that treat mathematics as the memorization of procedures, and that value the memorizers who can speedily regurgitate what they have memorized, turn slow, deep thinkers away from the subject. Even those who are successful develop an impoverished relationship with mathematics. When people encounter knowledge differently, doors open for them into a different world. They learn concepts that are compressed in the brain, and they build a solid foundation of understanding. They are able to welcome mathematical thinking into their tool kit and use it not only in math class, but in all subject areas. In our current system a few high-achieving students learn to think flexibly, and they go on to become trailblazers in different fields.

Teachers such as Marc Petrie, whom I discussed in the last chapter, and Nina Sudnick, in this chapter, were good teachers before they encountered the ideas I share in this book, and many students were successful in their classes.

But now that they approach math with depth and creativity and teach their students the importance of deep thinking, significantly more students are successful in the classes of both teachers.

Developing a good relationship with mathematics, as Bob Moses, one of the most influential leaders of the civil rights movement in the 1960s, says, is a civil right. It opens doors in school and in life. Many people believe that greater understanding and proficiency come about from greater knowledge and regard their task in learning to be the accumulation of knowledge. But research tells us that more effective and higher-achieving people engage in flexible thinking. Creative and flexible thinking, a much-needed and valued way of working, is actually inhibited by greater amounts of knowledge.[16] This is why when problems need creative solutions that involve seeing patterns and unexpected connections, it is often trained professionals who are unsuccessful and outsiders who solve the problems.

Adam Grant has written a book called *Originals: How Non-Conformists Move the World* in which he argues that we have long valued the rule-following, memorizing students. He notes that the students in the US often regarded as "prodigies"—the ones who "learn to read at age two, play Bach at four, breeze through calculus at six"—rarely go on to change the world. When scholars study the most influential people in history, they are rarely those regarded as "gifted" or "geniuses" in their childhood. Instead, the people who excel in school often "apply their extraordinary abilities in ordinary ways, mastering their jobs without questioning defaults, and without making waves." Grant concludes: "Although we rely on them to keep the world running smoothly, they keep us running on a treadmill."[17] Those who do go on to change the world are creative and flexible thinkers, people who think outside rather than inside the box.

Many people know that creative and flexible thinking is valuable, but they do not associate it with mathematics. Instead, they see math time as an area in which to follow rules and be compliant. But when we combine mathematics with creativity, openness, and out-of-the-box thinking, it is wonderfully liberating. This is something everybody deserves to know about and experience, and when they do, they do not look back.

The advantages of deep and flexible thinking apply to all subjects and avenues of life. We do not know what problems people will need to solve in the future, but they are likely to be problems we have never even dreamed of. Filling our minds with content that we can reproduce at speed is unlikely to help us solve the problems of the future; instead, training our minds to think deeply, creatively, and flexibly seems far more useful. The thinking of "trailblazers" whose brains were studied was found to be more flexible than that of regular people. They had learned to approach problems in different ways and not just rely on a memory. Speed and fixed approaches will only take us so far. In the education world and beyond, we must all challenge the assumptions about the benefits of speed and memorization and instead focus on flexible and creative learning. This will help us unlock our own and others' potential as learners.

6

A LIMITLESS APPROACH
TO COLLABORATION

THE FIRST FIVE KEYS that help unlock the limitless power of our learning—and living—potential draw upon knowledge of these ideas:

- brain plasticity and growth
- the positive impact of challenges and mistakes on our brains
- beliefs and mindsets
- a multidimensional approach to content that increases brain connectivity
- flexible thinking

All of these keys can unlock people, and sometimes one key alone can be incredibly liberating. If you believe that you cannot learn in a particular area or that only fast thinkers are capable, for example, then learning that these ideas are incorrect can free you up to pursue your chosen path. In this chapter, I will share a sixth key that works to unlock people, but it is one that can also be a product of being unlocked. The key centers on connecting with people and with a multitude of ideas. Connections and collaborations have a tremendous amount to offer the process of learning, and living.

LEARNING KEY #6

Connecting with people and ideas enhances
neural pathways and learning.

Why Is Collaboration Important?

Over my lifetime I have encountered a small number of fascinating situations, some through research and some through personal experience, in which collaboration and connection produced surprising outcomes. Some of these have related to learning, some to the pursuit of equity, and some to the advancement of ideas, even in the face of severe opposition. These different cases all shed light on something that neuroscience is also showing—when we connect with other people's ideas there are multiple benefits for our brains and for our lives.

Uri Treisman, a mathematician at the University of Texas at Austin, used to teach at the University of California, Berkeley. While Uri was at Berkeley, he noticed that 60 percent of African American students who took calculus were failing the class. This caused many to drop out of college altogether. Uri began looking at more university data and saw that no Chinese American students were failing calculus, so he asked the question: What is the difference between these two cultural groups that seems to be causing this discrepancy?

Uri at first asked the other mathematics faculty what

they thought the reason was. They came up with a range of reasons: perhaps African American students came into the university with lower math scores or an insufficient mathematical background; perhaps they were from less wealthy homes. None of these suggested reasons were correct. What Uri found, through studying the students at work, was that there was one difference—the African American students worked on math problems by themselves, whereas the Chinese American students worked collaboratively. The Chinese American students worked on their assigned math problems in their dormitories and in the dining halls, thinking about them together. By contrast the African American students worked alone in their dormitory rooms and when they struggled on problems, they decided they were just not "math people" and gave up.

Uri and his team set up workshops for the more vulnerable students, including students of color. They created what Uri describes as a "challenging yet emotionally supportive academic environment."[1] In the workshops students worked on math problems together, considering together what it would take to achieve at the highest levels on different problems. The academic improvement that resulted from the workshops was significant. Within two years, the failure rate of African American students had dropped to zero, and the African American and Latino students who attended the workshops were outperforming their white and Asian classmates. This was an impressive result, and Uri has continued this approach at Austin. His approach has now been used in over two hundred different institutions of higher education. In writing about the experience, Uri says:

We were able to convince the students in our orientation that success in college would require them to work with their peers, to create for themselves a community based on shared intellectual interests and common professional aims. However, it took some work to teach them how to work together. After that it was really rather elementary pedagogy.[2]

The fact that it took work to teach students how to collaborate with each other after they had spent thirteen years in school speaks to the problems in our school system, where the common pattern is that teachers lecture and students work through problems alone. The team leading the workshops was right to point out that success in college requires working with others and making good connections. Many people know this, but they still see no role for collaboration in learning. When Uri and his team encouraged students to work together, their mathematical learning paths changed and they found success. This success story was about learning calculus in college, but we could substitute any other subject and expect similar results.

Part of the reason students give up on learning is because they find it difficult and think they are alone in their struggle. An important change takes place when students work together and discover that everybody finds some or all of the work difficult. This is a critical moment for students, and one that helps them know that for everyone learning is a process and that obstacles are common.

Another reason that students' learning pathways change is because they receive an opportunity to connect ideas. Connecting with another person's idea both requires and develops a higher level of understanding. When students work together (learning math, science, languages, English—anything), they get opportunities to make connections between ideas, which is inherently valuable for them.

A similarly noteworthy finding came from the results of a large-scale testing program. In 2012, PISA assessments (international tests given to fifteen-year-olds worldwide, as mentioned earlier) showed that boys achieved at higher levels than girls in mathematics in thirty-eight countries.[3] This result was disappointing and surprising. In the US and in most other countries, the achievement of girls and boys in school is equal. This reminded me again of the ways that tests distort what students actually know and can do.

This was underscored when the PISA team issued a report showing that when anxiety was factored into the analysis, the gap in achievement between girls and boys was fully explained by the lower confidence of girls.[4] What appeared to be a gender difference in mathematics achievement was in reality a difference in mathematics confidence levels. Girls became more anxious when they took the individual math tests, a phenomenon that is well established,[5] and one that should make any educator pause before basing decisions on test performance.

The impact of the different testing conditions as well as the potential of collaboration for reducing inequalities were also shown by another assessment the PISA team conducted. In addition to the usual individual mathematics test, they did an assessment of collaborative problem solving. In this assessment students did not collaborate with other students but with a computer agent. They had to take on the ideas of the agent and connect with and build upon them to collaboratively solve complex problems.[6] This, to me, gauges something much more valuable than what a student produces on an individual math test. Instead of reproducing knowledge individually, students are asked to consider another's ideas and work with them to solve a complex problem. This is also more consistent with the world of work students are being prepared for.

In the test of collaborative problem solving, administered in fifty-one countries, girls outperformed boys in every country. This notable result was accompanied by two others—there were no significant differences in outcomes between advantaged and disadvantaged students, a rare and important finding. And in some countries diversity boosted performance. The team found that in some countries "non-immigrant" students achieved at higher levels when they were in schools with larger numbers of "immigrant" students, a fantastic result, suggesting that diverse communities of learners help students become better collaborators.

The results of the PISA assessment of collaborative problem solving shine a light on the pursuit of equity, revealing also the discriminatory nature of individual testing, something that anyone who gets anxious about high-stakes testing fully understands. What does it mean that for girls collaboration, even with a computer agent, increases their confidence levels and causes them to achieve at higher levels? Similarly, what does it mean that African American students go from failing calculus to outperforming other, previously more successful, students when they collaborate? This research reveals the potential of collaboration, not only for girls or students of color, but for all learners and thinkers. When you connect with someone else's ideas, you enhance your brain, your understanding, and your perspective.

Neuroscientists also know the importance of collaboration. Research shows that when people collaborate, the medial orbitofrontal cortex and the frontoparietal network are activated, the latter of which aids in the development of executive functions.[7] Neuroscientists refer to these different brain areas as the "social brain." When we collaborate, our brains are charged with the complex task of making sense of another's thinking and learning to interact. Social cognition is the topic of much current neuroscientific investigation.

Collaboration is vital for learning, for college success, for brain development, and for creating equitable outcomes. Beyond all of this, it is beneficial to establish interpersonal connections, especially in times of conflict and need.

Victor and Mildred Goertzel studied seven hundred people who had made huge contributions to society, choosing those who had been the subject of at least two biographies, people such as Marie Curie and Henry Ford. They found, incredibly, that less than 15 percent of the famous men and women had been raised in supportive families; 75 percent had grown up in families with severe problems such as "poverty, abuse, absent parents, alcoholism, serious illness," and other major issues.[8] Their study was conducted in the 1960s. Clinical psychologist Meg Jay, in her interesting *Wall Street Journal* article on resilience, reports that similar results would be found today and cites Oprah Winfrey, Howard Schultz, and LeBron James as examples of people who grew up in extreme hardship.[9]

Jay has studied resilience over many years and points out that people who survive hardship often do better, but not through "bouncing back," as some think, because the recovery process takes time and is more of a battle than a bounce. She also points out those who ultimately benefit from hardship, becoming stronger and resilient, do so when they maintain self-belief, when they "own the fighter within," and when they connect with other people. The thing that people who overcome hardship and do not become defeated by it have in common is that in times of need they all reached out to someone—a friend, a family member, or a colleague—and those connections helped them survive and develop strength.

The Power of Collaboration: Two Cases

I found myself on the wrong side of a very unpleasant disagreement and was able to overcome it because of collaborations. The saga began when I moved from King's College London to Stanford. I had just completed my doctorate and had conducted a very careful and detailed study of two schools that had demographically similar student populations but taught mathematics very differently. The study won the award that year for the best doctoral study in education in the UK. The book about the study also won an award for the best book in education.

I chose to follow an entire grade level of students over three years, from when the students were thirteen to when they were sixteen. I conducted over three hundred hours of lesson observations, watching the students work on mathematics. I interviewed teachers and students every year of the three-year study. And I also gave students applied mathematics problems as assessments and carefully analyzed the students' scores and their approaches to questions on the UK national examinations. The results were enlightening and were reported in newspapers across England. At the school that taught through traditional methods, used by most schools in England (and the US and many other countries in the world), where a teacher explained methods and students worked through closed-textbook questions, the students mainly disliked math and scored at significantly lower levels on the national examination than the students who learned mathematics through open, applied projects.[10]

At the school that taught students through projects, most of which took a few lessons to complete and required students to use and apply methods in different ways, the students enjoyed mathematics more and went on to score at significantly higher levels on the national exam.[11] The students from the project school outperformed the students learning traditionally, because they approached each question as an opportunity to think and apply various methods, whereas the traditionally taught students approached the questions by trying to recall information from their memory. In addition, the traditional approach maintained inequities between girls and boys and between students from different social classes. Those inequities that were present when the students started at the project-based school were eliminated over the three years of the study.

In a follow-up study I worked with a group of students from each school, now adults of about twenty-four years of age, who had scored at equivalent levels on the national examination. This showed that the students who attended the project school were in more professional and higher-paying jobs.[12] The adults reported using their school mathematics approach at work—asking questions, applying and adapting methods, and being more proactive in changing their job if they didn't like it or in applying for promotions. The adults from the traditional school said that they never used any of the mathematics they had learned in school, and they seemed to have taken the passive approach they were required to adopt at school into their lives.

When I presented the results of the study at a conference in Athens, Greece, in the summer following my PhD completion, I was approached by the dean of Stanford's Graduate School of Education and the chair of the mathematics search committee. Both had been in the audience, and they told me they had been looking for a new mathematics education professor. They asked me to consider working at Stanford. At the time, I was very happy as a researcher and lecturer at King's College and told them I was not interested. But over the next few months, they sent me picture books of California and persuaded me to come for an interview, to experience Stanford and California. They knew what they were doing, and when I eventually agreed to spend some days on the California coast, I became enchanted. Later that year, I moved to Stanford.

A few months into my time there, I received an email from a James Milgram in the math department saying he wanted to meet with me. I did not know much about him and agreed to meet him at his office in the math department. It was a disturbing meeting, during which he told me that teachers in the US do not understand math and that it would be dangerous for me to publicize my research evidence in the US. I, of course, countered this idea, but he was not interested.

Over the next few years, I received a National Science

Foundation presidential award; these are given to people considered to be the most promising researchers in STEM disciplines. This provided funding for a US study similar to my UK study. For the new study, a team of graduate students and I followed approximately seven hundred students through four years at three different high schools that taught with different approaches.

The study produced results similar to those of the UK study. The students who learned mathematics actively, using and applying different methods for complex problems, achieved at significantly higher levels than those who reproduced methods a teacher had rehearsed. Again, the students also developed significantly different ideas about mathematics, and the students who learned more actively were ten times more willing to continue with mathematics beyond high school.[13] Among those who learned mathematics passively, watching a teacher work through problems, even high-achieving students told us they could not wait to drop math and carve out a future that did not include any more math courses.

When the results of this study started to emerge, Milgram accused me of scientific misconduct. This is a very serious charge that Stanford had to investigate by law and that could have ended my career. I was required to give a group of senior Stanford professors all of the data we had collected over the last four to five years. Stanford investigated Milgram's claim, found no evidence at all that there was any cause to question our results, and ended the investigation. But Milgram was not finished, and his next move was to write a collection of academic differences that I regarded as savage criticism and publish them online. I decided at first, on Stanford's advice, to ignore this. I was, however, unimpressed by the whole set of events and decided to move back to England.

I was awarded a prestigious Marie Curie fellowship that funded my work for the next three years at Sussex University. I hoped that a new coastal environment would help obliterate the memories of the last few months and give my two daughters (ages six months and four years) a good place to grow up. But over the next three years I realized people were reading and believing Milgram's claims.

Milgram was not alone in his mission to stop reforms in school mathematics, and he and others sought to undermine my work on websites, saying I made up data and the schools in my UK study "only existed in my mind." On a site they thought was private, available only to the people working to stop reforms, one of them wrote, "This is the worst possible scenario, a researcher in a top university, with data." The professor who had been my PhD advisor, Paul Black, an incredible scientist who has been knighted by the pope for his services to education, was appalled by the criticism from American professors and wrote to them, but that also made no difference.

The Graduate School of Education at Stanford regularly asked me to come back to my position at Stanford, which had not been filled, and one cold February day, three years after I had left California, I started to consider it. I had battled through the pouring rain that dark morning to take my daughters to the local primary school. When I returned home and dried off, I opened my laptop to find an email from one of my former Stanford colleagues, asking again if I would return. Perhaps it was the cold weather or the rain, I am not sure, but for the first time I thought to myself, "Maybe I should go back." At the same time I also promised myself that I would only go back if I fought to stop the campaign of slander that had been waged against me.

A few months later I was back in my position at Stanford. Many people assumed I had moved back in order to swap the gray UK weather for the sunny blue skies of California. This may have been part of it, but what I really missed when I lived in the UK was the warmth of people in California, and the US more generally. During my years in the US, I had been made to feel by many teachers that my work was really helping them.

Fortunately, the school of education at Stanford had by now appointed a new dean—an amazing man called Claude Steele, who has pioneered work in stereotype threat. He looked at the details of what Milgram and his friends had written about me and their other writing and immediately recognized the kinds of people we were dealing with. Together we decided on a fairly simple strategy—we made a plan that I would write down the details of their conduct and publish them.

I remember the Friday evening clearly. The other education faculty were gathering for a party. I stayed at home and clicked the button that made public my new webpage detailing the behavior of the men.[14] That was when everything changed. That night I also joined Twitter, and my first post linked to the details of the academic furore. It spread like wildfire, and over the weekend my webpage was the most tweeted story in education. Within forty-eight hours I had been contacted by reporters from across the US, who ran news stories detailing the events.

Then something else happened. I started receiving emails from other women professors and scientists, all sharing their own negative experiences. Within a few days I had received about a hundred emails, all of them sympathetic and most of them from other women detailing stories of their own ideas being put down or rejected by men in universities. The email collection was a clear indictment of university culture and a sign that we are far from achieving gender equality in higher education. I am sure everyone reading this book would have thought that by 2013 university departments would not still be discriminating against women, but it was clear from reading the emails that there are still plenty of men in positions of power who do not think women belong in STEM fields. They probably do not realize the extent of their discriminatory ideas and would be surprised by my making that statement, but their actions to suppress the work of women, detailed in many of the letters I read, revealed the discrimination clearly.

As the weeks and months went by after the details of their behavior became known, I started to feel what I can only describe as warmth—the warmth of support from tens of thousands of teachers, mathematicians, scientists, and others. The walls I had constructed inside me started to melt, and I became more open.

At the same time, we were entering a new era in education, with initiatives led by President Obama and a wider awareness that change was needed. I left behind the trauma that the saga had borne and became inspired to share the evidence more widely—in online courses and on youcubed. As I write this, it has been more than six years since the night that I revealed my experiences, and during that time we have had millions of hits, downloads, follows, and approximately one-half of schools in the US are using our lessons and materials. Ironically, part of the reason that we have such a huge following is because people saw me standing up to those challenging me.

Sharing my story started a process of connecting that was, for me, transformative. Before going public I had walked alone with the burden of the damaging experiences—but afterward people started to reach out to me. The support of other people started a noticeable change within me. Perhaps I am more aware of the change because I had been hurt so badly, was so closed inside, and then went through a transition period. The phrase "What doesn't kill you makes you stronger" rang true for me, as I developed a strength that I know came from the episode.

The connections that I formed through the process of sharing my story and experience helped me and helped the people I connected with. Through connecting with others I learned how I could become unlocked and open rather than closed.

People still come after me, especially those helped by the anonymity and distance of social media. They think they can throw out insults and write abusive words about a woman working to improve education, but I am a much stronger person now. An idea that I hold on to when I read aggressive attacks is this: "If you are not getting pushback, you are probably not being disruptive enough."

Education is a system in which we need to challenge the status quo because it has failed so many. So when I offer alternatives to what has always been assumed to be best and people come after me, I am able to ignore their aggression, knowing that they are lashing out because what I have suggested has affected them in some way. I have learned to approach pushback differently—instead of letting it get me down or causing me to doubt myself, I view it as an opportunity.

This is a valuable mindset change. If you try to make productive change or suggest something new, in learning or in the workplace, and people ridicule or dismiss you, try to view their criticism as a sign that you are making a difference. Pushback is a positive sign; it means that the ideas that are ruffling people's feathers are powerful. The showman Phineas T. Barnum, founder of the Barnum & Bailey Circus (also the subject of the movie *The Greatest Showman*, starring Hugh Jackman), once said: "Whoever made a difference by being like everyone else?"

I love that quote. It helps me know that new ideas will never be easy for some people to accept, yet they are very important. The ideas that are hardest for people to accept are those that go against the status quo, and they may be the most important of all. When I teach about new learning methods based on evidence from neuroplasticity, I tell my listeners that they too may experience pushback when they share the new evidence with others. People are so invested in the idea that intelligence and learning ability are genetically determined that they resist any notion to the contrary, particularly if they are the ones who benefited from their firmly held belief.

When people ask me how I coped with the events I have described, I am very clear that there was one action that changed everything for me. It was the sharing of my experience with others and the incredible response of educators and scientists across the world who contacted me. It was connections with people—some of them in person, many of them online—that healed the wounds I was hiding. When people ask me what they should do when their work is attacked, I always suggest that they find people to connect with. Doing this online does not work for everyone, though that method is certainly more available now than ever before. For some it is better to reach out to a colleague or family member. But however it is done, that connection with another person or persons will be invaluable.

Before I turn to the question of how we, as parents, educators, and managers, encourage a limitless approach to connections and collaborations, I would like to relate one more story of collaboration that started in a high school and is now a global movement.

Shane was beginning high school when he hit an all-time low. He had just started at a typical large US high school with high expectations for his experience there, but within weeks he said it was the "loneliest experience he had ever felt." Shane, in a powerful video that has had tens of thousands of views,[15] describes feeling like an outsider, someone who didn't belong. It was this deep-seated feeling of emptiness that led Shane to an appointment with his guidance counselor. Shane agreed to this because he thought it might result in his transferring to another school.

Instead, he walked out of the meeting with the recommendation to join five different clubs at the school. Shane was skeptical at first, but he joined the clubs and started to notice some things shifting. He had people to say hello to in the corridors, and the more he got involved in school life, the more he felt he was part of the school community. Shane discovered that the more he did, the better he felt about himself; the more involved he got, the more "connected, driven, and motivated" he felt. He now reflects that he felt like an outsider because he was one; the only thing that changed was that he put himself on the inside—and that shifted everything. This was such a powerful change that Shane became inspired to share his experience with others and to start what is now a global movement—to help young people to become more personally connected with others.

Shane initially had an idea to hold an assembly at his school to help other students know what can happen when they connect with others and then match them with clubs of interest to them. They expected about fifty students, but word spread and four hundred students from seven different schools attended the assembly. The following year that number grew to a thousand, and the numbers have continued to grow every year. Shane started the movement "Count Me In," which has now impacted more than ten million people, with speaking programs that have reached students in over one hundred countries. When I interviewed Shane for this book, he highlighted the challenges today's young people face in forming meaningful connections:

Teenagers today have it harder than any other generation by far, in my educated opinion. Not only are they dealing with all the same issues we've seen for generations, but also things like peer pressure, bullying, social isolation that can really be harmful on your upbringing and the trajectory of your life. These are now 24/7 issues for every kid because of technology and smartphones and how much they are plugged in online, yet unplugged in reality and in community. Those community connections, I think, are the key in forging something—so we can see the world just differently enough that we begin to feel a greater sense of self-acceptance and belonging.

He makes a crucial point, and his movement to create greater connections among young people serves a much-needed purpose, as he stated in our interview:

The more you get involved, the more you immerse yourself in the community, the more connected you feel, and the more different you see things, the stronger that lens becomes, and the softer things become. The real defining moment for me that I can pinpoint is when I started living from this place of: My life is bigger than this moment, and it doesn't matter what's going on, how dark or desperate I feel inside. I know for a fact, with absolute conviction and certainty, that my life is bigger than this moment, than any one moment.

Shane's movement has been particularly helpful for young people who feel isolated, who are going through difficult times at home, or who are facing any of the myriad issues that impact young people. He reflected that the main response that differentiates those who change positively from those who don't is their perspective, or their mindset. Shane's movement is also a helpful reminder that even—or perhaps particularly—in a world of online connectivity, genuine human connections are something that everybody needs and that changes people's lives. Shane found that they helped young people know that their lives are bigger than the moments they are in now and that no matter how hard a situation may be, connections with people bring you out of it.

A Limitless Approach to Connections and Collaborations

I have related studies and examples of situations in which collaboration has completely changed students' achievement and people's lives more broadly. But what have these ideas got to do with the keys I have presented in this book—about brain growth, struggle, and multidimensionality, for example? My work over recent years and the interviews I conducted for this book show me that there is a different way to collaborate and connect—a limitless way—and when we teach a limitless approach to students and others, they find that connections, meetings, and group work are infinitely more productive, enjoyable, and generative. In the remainder of this chapter, we will look at the limitless approach to collaboration as well as some strategies for bringing it about in classrooms, homes, and workplaces.

Teachers know that group work can be really hard—especially when students hold negative ideas about each other's potential and status differences emerge. For teachers

who know the importance of students talking to one another and connecting with each other's ideas, this presents
a dilemma. Parents face similar difficulties when they see
siblings fail to interact well—coming into conflict instead
of sharing their thinking and ideas productively. The difference between positive and negative interactions frequently
depends on three actions that teachers, parents, and managers can move toward: (1) opening minds, (2) opening content, and (3) embracing uncertainty.

(1) Opening Minds

To interact well, people need to have an open mind, and to
develop it, they need to learn to value difference. Students
will start to appreciate and think positively about each
other, if teachers highlight the importance of different ways
of thinking—about math, history, science, anything. Many
teachers lament the fact that students do not interact well in
groups, but this is largely because, having closed minds, they
think they are looking for one idea, one answer, so difference
and diversity are not valued. When we change this perspective, for children or adults, it changes how they interact with
others, in classrooms and in life.

A few years ago, I conducted a four-year study of students
moving through various high schools. In one of the schools
the students were taught to interact well in groups. They
were taught to listen to and respect each other, and they were
taught that the different ideas they each shared were valuable. Something fascinating happened in that school that I
have referred to as the creation of "relational equity."[16]

Usually people think of equity in terms of test scores: Are
all students scoring at roughly the same levels? But I have

suggested that there is a more important form of equity—one that concerns students' learning to interact well and value each other. I contend, as have others,[17] that one of the goals of schools should be to produce citizens who treat each other with respect, who value the contributions of others with whom they interact, irrespective of race, class, or gender, and who act with a sense of justice, considering the needs of others in society. A first step toward producing citizens who act in such ways must be the creation of classrooms in which students learn to act in such ways, for we know that students learn a lot more than subject knowledge in their school classrooms.

I extend the notion of equity to the relations between students with the assumption that the ways students learn to treat and respect each other will impact the opportunities they extend to others in their lives in and beyond school. The teachers in the school I studied had worked to create an environment in which students respected each other by beginning with content—the students were used to sharing their different ideas on the content being discussed, as evidenced by these students' comments. The interviewer asked, "What do you guys think it takes to be successful in math?" The students responded:

> *Being able to work with other people.*
>
> *Be open-minded, listening to everybody's ideas.*
>
> *You have to hear other people's opinions, 'cause you might be wrong.*

Parents of high-achieving students sometimes complain that their student is being used to educate others; their student could just as easily work alone and zoom through material. But these students were taught that part of being a community, such as a class of learners, involves looking out for each other. The students developed an important perspective on their responsibility for others. One student said:

> I feel it is a responsibility, because if you know something you have to, and somebody doesn't know, I feel it's your right, you have to teach them how to do it. Because it's only fair to them that they get as much out of it as you're getting out of it, because you're both in the same classroom.

Despite the fears of some parents of high-achieving students, the students in this approach who improved the most in math achievement were the highest achievers.[18] They increased their achievement more than other students at their school and more than the highest achievers in the other, traditional schools we studied. Their achievement gain came from their time explaining work to others—which is one of the best opportunities for students to understand more deeply themselves. An important part of the respect students developed for each other came from teaching them to see content in a more open way and teaching them to value difference. Just as information about mindsets teaches students to move from fixed to growth thinking, teaching the respect of varying ideas about content leads to a valuing of difference and diversity in other areas. Valuing both growth and difference is a powerful way of opening minds.

Holly Compton, a teacher I introduced in Chapter 4, had told her fifth-grade students: "Everybody has a different way of approaching things and you can always learn and grow." She told me that these ideas had caused students to be less egocentric. In their interactions now, instead of insisting on their way of thinking or working and closing down because others have a different idea, students think: "Oh, you know what? This is how I'm thinking about it, but I know others think about it another way." This acceptance of different ways of thinking has led to greater tolerance and appreciation of each other. As Holly reflected:

> They know that other people have good ideas too and they also know that they should open their mind to hearing other people's solutions, because that might be a new idea for them that they hadn't yet thought of. And so that mindset of, "Hey, maybe your idea is something that I could add to my idea" is a huge one for kids.

Many reformers in education who work to change student experiences in classrooms work on content, finding new ways to approach topics, often with cool technology. But imagine what students' learning and lives outside school would be like if they learned to collaborate with others more productively, going into conversations with an openness to hear and understand what others have to say. This would change classroom dynamics as well as many other aspects of students' lives.

Holly reported on one interaction she had overheard:

> I was in a classroom today where the kids were challenging each other's thinking, and one kid was disagreeing with another, but he said, "I think I know what you were thinking. I think you were thinking"—and he explained—"but it's really this." And the other kid said, "Oh yeah, that is what I was thinking." And these were first graders! They're not

supposed to be able to take the perspective of others when they're that little.

The first graders had learned about mindset and multi-dimensionality, and it had opened them to considering other people's perspectives, which started different—limitless—pathways for them.

In Chapter 3 I noted the research showing that people who develop a growth mindset become less aggressive toward others. Interestingly, this stems from changing the way people feel about themselves. According to the research, people with a fixed mindset had thought they themselves could not change and so felt more shame about their own actions, which probably caused them to lash out more. When they learned that nothing was fixed and any change was possible, they felt less shame about themselves and started to see others in a different way. They stopped seeing others, even adversaries, as inherently bad people and instead saw them as people who had made bad choices but could change. This caused aggressive tendencies to fade away and to be replaced by forgiveness.

These profound changes in people come from an opening of perspective, an opening of minds. We are probably only at the beginning of understanding how a shift in belief, from resisting difference and potential growth to embracing them, impacts the ways in which people interact in the world. We have seen that changed beliefs improve learning and health and decrease times of conflict. If students and children enter into collaborations with the belief that anyone can change and grow and that different ideas should be valued, it will dramatically change the interactions that follow.

(2) Opening Content

I have described the importance of opening minds, so that students and others appreciate difference and diversity. An important way for students to learn this perspective comes about when the content of academic subjects is taught in an open way. Similarly, when people in business settings are encouraged to be appreciative of multiple opinions and viewpoints, they start seeing themselves and regarding others differently.

I first began thinking more deeply about the connections between mindset, treating content in an open way, and ways that people interact when we taught our summer camp to eighty-three students a few years ago. During the summer we had observed productive group work, in which students respectfully shared ideas with each other. This group work contributed to the students' learning and high achievement, as they helped each other and discussed ideas well.

The students told interviewers that group work did not work well in school, but at our camp it did. The interviewers asked what the difference was. The students explained that in school one person would do all the work, while the others talked about clothes. But at our camp, group work was started by going around the group and having the students ask each other: "How do you see it? How would you approach it?" When students start group work by sharing their perspective and way of seeing a problem, they become invested and feel included in the work, which is the perfect start to group interactions.

As the students at our camp learned that a multidimensional approach was a good way to learn math, they started to value each other's ways of seeing and solving mathematical problems. This caused them to value each other more and avoid the negative ideas that are often developed in classrooms—one of which is that some people are more worthwhile than others.

This simple strategy of asking people how they see something, how they interpret something, can be used in many different settings. If people started business meetings by asking for ideas and interpretations, with an openness to all ideas and no judgment or expectation of a particular answer, it would cause people to feel valued and included, which would change relationships and productivity. Any teacher, in any content area, could use this strategy to help spark students' thinking and increase their inclusion. Importantly, the connections that arise from more open beginnings lead to more worthwhile conversations and ultimately better relationships, thinking, and work. Later I will provide some strategies to help teachers treat content in an open way, encouraging different interpretations and ideas.

(3) Embracing Uncertainty

Across the sixty-two interviews I conducted for this book, I frequently encountered an idea that people said contributed to their becoming unlocked in interactions with others. Many spoke of the power of letting go of the belief that they always had to be right in conversations with others. This meant becoming comfortable with uncertainty. The new perspective stemmed from learning about the value of challenges and mistakes for our brains. When people learned that struggle was productive, it caused them to open up in different ways, one of which was relinquishing the notion that they had to go into every meeting as an expert.

One of those who talked about this was Jenny Morrill, who teaches mindfulness to her students and has written a book, with Paula Youmell, called *Weaving Healing Wisdom*.[19] In the book she shares her methods for focusing on the moment with students. When I interviewed Jenny, she also described an interesting change in herself. Despite Jenny's well-developed understanding of mindfulness, her own relationships changed significantly when she added in "the brain science piece."

Prior to learning about the value of struggle and brain growth, Jenny "felt like an island." She described to me a mindset that I am sure many share—of feeling that she had to be an expert when interacting with others, of being fearful of revealing a lack of knowledge. And as a teacher in a classroom, she felt she had to be the one who knew everything. But Jenny's perspective has changed, and she now embraces uncertainty and opens up more to her community of colleagues. Part of this change has involved letting go of the idea that she is being judged. Jenny described her new perspective:

> *Being willing to feel uncomfortable with not knowing something and still know that I don't have to give up on something just because I don't understand it right away. And I have other resources that I can utilize to increase my learning as an educator, as a person. So for me, it's just . . . I always felt like I was an island and I had to show up knowing. . . . I think for me, it's changed the way I navigate life in terms of I listen better, I think. I feel like I grow and learn by collaborating, so I think I've opened up a different way of connecting to my community of colleagues so that I can learn better, and sharing is really learning. That whole idea of*

letting go of judgment and knowing your worth changed me
as a person.

I don't know why Jenny felt that she had to be "an is-land" before, but her new, more open position—collaborating with others, listening to them, being vulnerable, and learning from them—is enhancing her life immeasurably. Jenny told me that she no longer sees herself as the sole expert in the room and instead encourages students to be leaders, and she recognizes that she can learn from her students as well as other adults. In taking this perspective, Jenny has brought about powerful shifts in her students as well as in her own collaborations with colleagues and friends.

One of the important changes that many of the inter-viewees describe is being more resourceful when they face roadblocks. Instead of pretending they know or understand something they do not, they search for resources. Jenny spoke of some of the resources she now uses:

> *And now I know that I can show up, and not necessarily*
> *know, but I can use my intuition, I can use my colleagues,*
> *I can google something, I can watch a video, I can watch a*
> *YouTube channel, a YouTube video about how to explain*
> *a math process or something or . . . I'm never gonna stop*
> *learning. Whereas before, I felt like I had to walk in the door*
> *knowing it. That was my fixed mindset. I had to look like*
> *everything was understood and handled and under control,*
> *and that isn't necessarily the way I do things now. So I've*
> *released that idea. . . . I don't respond to change with as*
> *much tension as I used to. So I'm more open to recognizing*
> *I'm experiencing something that might be uncomfortable for*

*the moment, but that I can learn how to navigate. The more
I relax with it, the more I can navigate it.*

This new approach—of embracing uncertainty instead
of pretending to know everything, of looking for resources
to learn more—seems to enhance people's connections with
each other as well as people's way of being in the world.

Approaching content with uncertainty and vulnerability
is a trait I also recommend to teachers I work with. When students see their teacher present correct content all the time,
always knowing the answer to any student question, always
being right, never making mistakes, and never struggling, it
creates a false image of what it means to be a good learner,
in any subject. Teachers should embrace uncertainty and be
open about not knowing something or making a mistake.

If you are a teacher, share these times with students so
that they know such times are an important part of having
expertise. When I teach my undergraduates at Stanford, I
give them open mathematics problems to explore. They take
them in all sorts of directions, some of which are new to me.
I embrace these moments and admit that I do not know, saying, "How interesting. I have not seen that before. Let's explore it together."

Sharing uncertainty is an important strategy for learners, managers, teachers, and parents. You will find that when
you are vulnerable and admit not understanding a particular
point, others join in, and soon everyone is sharing openly and
productively. If you are a parent, discuss ideas with your children not as the expert in the room, but as a thought partner.
Ask your children to teach you things—it will be something
they enjoy; it will give them pride and enhance their learning.

Admit to your children when you do not know something, but you have an idea for a way of finding out. Never pretend you know something that you do not. It is much better to model a mindset of discovery, of finding out, of being curious, and of being happy living in a place of uncertainty—because that allows you to find out something new. I sometimes tell my students at Stanford that I do not know what to do next in a mathematical problem and ask them to show me. They always enjoy this and learn a lot from it. They are learning that uncertainty coupled with a desire to learn is a good approach to take in any learning situation.

If you are a learner and don't have people to discuss ideas with, see if you can find a way to connect online. Join chat rooms and connect with people on social media, asking them any questions you have. A few months ago we invited followers of youcubed to a Facebook group. Now we have eighteen thousand people in the group. I love to see the ways that people openly ask questions of each other in the group—even questions teachers may be expected to know the answers to. Sometimes math teachers admit that they do not understand particular points in math—and twenty different people jump in to help them and discuss the ideas with them.

I always admire the people asking the questions, because they are showing that they are unlocked enough to be vulnerable and reach out to others. Other times people just share something they are working on and invite others to connect with the ideas, which I also love to see. Instead of viewing others you work with or learn with as competitors, start seeing them as collaborators, people you can be open with, people with whom you can form lasting connections. An open mind and the willingness to embrace struggle and the need to explore many different viewpoints are key to this life-changing approach.

Strategies to Encourage Limitless Collaboration

When I teach students of any age—middle-school students or undergraduates—I use a number of strategies to encourage good communication, which can be used in companies or classrooms by teachers and by learners. First, I always conduct a thought exercise about people's likes and dislikes with regard to group work. This is the first step before asking any students to work together on a problem.

I ask students to sit in groups and discuss things they do not like people to do when they work together on a problem. The students always come up with interesting ideas. And it's really important that students get the chance to voice these things out loud, for example, "I don't like it when someone tells me the answer," "I don't like it when someone says 'This is easy,'" "I don't like it when people are working more quickly than me," or "I don't like it when people dismiss my ideas." I collect one idea from each group and keep moving around the room until I've covered all the groups. I write the ideas I've collected on a poster.

Then I ask the students to share with their groups things they do like people to do when they are working together. They come up with comments like, "I like it when people ask me questions instead of showing me the way to do something," "I like it when we all start by sharing our ideas," and "I like it when others listen to my ideas." I collect one idea from each group and write them on another poster. I tell my students that I will keep both posters up in the classroom as reminders for our group work during the term.

A second strategy I use is one that I learned from educator and friend Cathy Humphreys, which she learned from mathematics educators in England. I use this in my math classes, but it could be used in any content area. It involves an approach that teaches students to reason well. Reasoning—setting out different ideas, providing justification for the ideas, and explaining the connections between them—is important in just about any subject area. Scientists often prove theories by finding cases that work or disprove them with cases that do not, but mathematicians prove things by reasoning.

I teach students that it is important to reason well—setting out ideas and the connections between them. I also teach them that it is important to convince others and that there are three levels of doing so. The easiest or lowest level is convincing yourself of something, the next level is convincing a friend, and the highest level is convincing a skeptic.

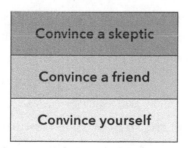

I also tell students that I want them to be skeptics with each other—to ask questions such as: "How do you know that works?" and "Can you prove it to me?" The students in our summer camp loved being skeptics with each other; they fully embraced the role, and the classrooms became filled with students asking questions and reasoning. We were particularly pleased about that, knowing that when you connect with someone else's idea in mathematics and other subjects, it both requires and creates a deeper level of understanding.

Collaboration can be extremely engaging for students, but I have met many students who hate group work. This is because they did not have good experiences working in groups. The groups were probably not set up well. They may have had problems that were closed, or students may not have been taught how to listen, be respectful, and have an open mind. But for the vast majority of students, subjects come alive when they are discussed and when students think about the different ways problems may be approached, why their ideas work, and how they are being used.

Many people believe that the essence of high-quality learning is working alone, studying hard. Artists' depictions of thinking and learning are often drawings and statues of individual people thinking hard. Rodin's *The Thinker* is one of the most famous examples—a man sitting with his chin rested on his fist, apparently deep in thought. But thinking is inherently social. Even when we read a book alone, we are interacting with another person's thoughts. It is very important, perhaps the essence of learning, when we develop the capacity to connect with another person's ideas, to build them into our thinking, and to take them forward into new areas.

When people embrace uncertainty, stop pretending to know everything, and instead look for resources to learn more, it opens up a different way of being in the world. This seems to be the essence of being limitless and is something I often observe in my Stanford colleagues. Some people, when faced with a challenge, give up, saying, for example, "I don't know how to use that software." Others approach the same challenge, saying, "I don't know that software, but I will learn it—I will look up videos and advice and teach myself. No worries." I meet these two approaches often and always note with admiration those who approach new challenges or learning with a limitless approach. They always end up achieving more and making use of a broader range of opportunities.

The pressure people feel to know everything is very real, but is something many people described letting go of when they learned about brain science and the value of struggle and openness. How often do people go into meetings or classrooms, worried that they do not know enough? And how refreshing would it be if people embraced times of not knowing and became more willing to acknowledge struggle? If people let go of the charade that they know everything and instead embraced uncertainty, it would give them a different way of navigating through the day, enabling more productive interactions. If people approached conversations in this way, it would help companies operate much more smoothly, friendships would develop better, and people would relax into their work and function more effectively.

Mathematics is often depicted as the most solitary of subjects, but it is a discipline, like all others, that has been built through connections between ideas. New ideas and directions come from people reasoning with each other, setting out ideas, and considering the ways they are connected to each other. Parents, particularly of high-achieving students, often say to me, "My child can work out the answers correctly. Why should she have to explain them?" But such parents are missing an important point—mathematics is all about communication and reasoning.

Conrad Wolfram, well known for his work with Wolfram Alpha, the online computational knowledge site, and director of Wolfram Research, told me that people who are unable to communicate their mathematical thinking and ideas are of no use to him as employees, because they cannot take part in team problem solving. In team problem solving, when people communicate their thinking, others can connect with their ideas. Critical evaluation by many minds also guards against incorrect or irrelevant ideas. When people cannot communicate an idea or come up with the reasoning that led them to it, they are not particularly useful in a team of problem solvers. I am sure this principle is true of all areas—people who can explain and communicate their ideas to others, whether in math, science, art, history, or any other area, are more effective problem solvers and are able to make a larger contribution to the work in companies and other groups.

The six keys that have been presented play an important role in changing people's communication and, consequently, create a multitude of life opportunities. Many people are too locked up to be good communicators. They are fearful of saying something wrong, and they worry that what they say is an indication of their worth, that they are being judged by others. When people learn about mindset, brain growth, multidimensionality, and struggle, it often unlocks them, gives them a limitless perspective, and enables them to let go of the fear of being judged. Instead, they embrace openness and uncertainty and become more willing to share ideas, which, in collaboration with others, grow into solution pathways. Such collaborations enhance people's lives, and the very best collaborations, it seems, start with a limitless approach to people and to ideas.

LIVING WITHOUT LIMITS

W E ARE ALL learning all the time. Schools and colleges may be the places we associate with learning, but they are not the only places in which learning happens. Our lives are filled with moments of learning, and it is in the myriad of opportunities to connect differently, with ideas and with people, that the learning keys in this book can come into play. My aim in writing this book is to equip you with ideas that can enhance all of your interactions, so you can live life as fully as can be and help unlock others by sharing the information you learned here.

The Swiss scholar Etienne Wenger created an important framework to help people think about learning differently. He states that when we learn something, it is more than just acquiring knowledge or accumulating facts and information, because learning changes us as people.[1] When we learn new ideas, we see the world differently—we have a different way of thinking and a different way of interpreting every event in our lives. As Wenger says, learning is a process of identity formation. Psychologists used to see identity as a static concept, maintaining that we all have "an" identity that we develop as children and keep throughout our lives. But more recent work has given identity a more fluid meaning, suggesting that we can all have different identities in different parts of our lives. You might, for example, present yourself differently as a member of a sports team than you do in your job or in a family role. I wrote this book because I know that when we learn about brain growth, mindset, and multidimensional deep and collaborative thinking, it unlocks aspects of our true selves. These ideas do not turn us into different people, but they can set free what was in us already, what was always possible for us but in many cases not being realized.

In preparation for this book, my team and I interviewed sixty-two people whose ages range from twenty-three to sixty-two and who come from six different countries. I asked, on Twitter, that people contact me if they felt they had been changed by learning about brain science, mindset, and the other ideas I have talked about. My team and I interviewed the people who replied over a period of several months. The interviews surprised me. I had expected to hear interesting stories of the ways people's perceptions and ways of thinking had changed, basically that they had adopted a different mindset. What I heard went well beyond that. The people we talked to described the *many* ways they had changed—in their connections with others, their ways of approaching new ideas and learning, the ways they parented, and the ways they interacted with the world.

The first initiative for change was learning about brain growth—the science of neuroplasticity. Many people we interviewed had accepted the notion that there were limits to their ability and that they couldn't do certain things. Because of my field and my Twitter followers, many of those we interviewed were talking about the ability to learn math, but they could have been talking about any subject or skill in their lives. And when they discovered that they could, after all, learn math, they went on to change their ideas about learning in general, realizing that anything was possible.

Angela Duckworth introduced the world to the concept of grit—of being dogged and determined in pursuing an idea in a particular direction.[2] Grit can be a really important quality, but it necessitates a narrow perspective, a zeroing in on the one thing that can bring about success. Athletes who achieve world-class success have all had to have grit, focusing on one activity and letting others go, and Duckworth herself talks about the need to be selective and not broad. This can work for some people, but is not the best approach for everyone. I know of people who have followed one pathway, turning down everything else, but then not quite reached their dream and been left stranded, unable to return to the other pathways that were once options for them. The idea of grit also has an individual focus, whereas scholars point out that more equitable outcomes often come from the work of community.[3] When young people excel and break through barriers to do so, it is rare for the achievement to be an individual accomplishment and much more typical for the achievement to be a collective effort, coming from teachers, parents, friends, other family members, and community allies. Grit does not capture this important feature of achievement, and may even give people the idea that they have to go it alone—and achieve through their own focused determination.

Being limitless is different from having grit. It is about a freedom of mind and body, a way of approaching life with creativity and flexibility, which I believe is helpful for everyone. People who approach life with a limitless perspective are

also dogged and determined, but they are not necessarily focused on one pathway. Freedom and creativity can lead to
grit, but grit does not lead to freedom or creativity.

During the writing of this conclusion, I became aware
of an incredible young man in England named Henry Fraser. I first read his book, *The Little Big Things*,[4] with a fore-
word by *Harry Potter* creator J. K. Rowling, and then I got in
touch with Henry. The book details Henry's life-changing
accident, when he became paralyzed after diving into the
sea in Portugal.

Henry had just completed his penultimate year of high
school; he was an avid sports enthusiast and played on the
school rugby team. At the end of the school year, he accepted
an invitation to travel to Portugal with a group of his rugby
friends for a few days in the sun and a well-earned break. On
the fifth day of his vacation, he dove under the water, hit his
head on the seabed, and severely crushed his spinal cord, an
injury that left him unable to use his arms or legs.

The next few days were spent in ambulances and operat-
ing rooms, with his parents, who had anxiously flown across
Europe, by his side. Early on in his recovery, Henry hit what
many would expect—a huge low, when one day he caught a
reflection of himself sitting paralyzed in a wheelchair. He
broke down, suddenly realizing how his life had changed
forever.

In the early days of Henry's recovery, he spent five weeks
without seeing the outside of his hospital room and six weeks
not being allowed to eat or drink. But when Henry started to
recover, he developed a new and life-altering mindset. When
he first felt sunlight again, immense gratitude welled up inside. When he breathed fresh air, he felt extremely happy, and

when he read the words on the cards sent to him by well-wishers, he felt humbled and grateful. These feelings did not stop for Henry, who made an active choice to move beyond regret and instead choose gratitude. For many in Henry's situation, the last thing they would feel was grateful, but Henry, who remains paralyzed from the neck down, does—approaching every day with happiness, embracing the opportunities he has to learn. Henry named one of his book chapters "Defeat Is Optional," and his mindset through adversity is one we can all learn a great deal from.

After Henry adapted to his life in a wheelchair, he set about learning to paint—holding the brush in his mouth. Now his stunning art is shown in exhibitions across England, and Henry's bestselling book is inspiring readers across the world. How did Henry achieve all of this, after his life changed so dramatically one summer? Instead of sinking into despondency, Henry has become extraordinarily accomplished and now inspires millions. A large part of this came about for Henry because he believed that if he tried, he could achieve.

When he hit rock bottom that day, seeing himself in the wheelchair, he probably felt locked up and severely limited, and he had good reason to. He chose instead to unlock himself and break out of the limits, with positive beliefs about his potential and his life. At the end of his book, Henry says he is often asked about the days when he "must" feel down and ask, "Why me?" This is his beautiful response:

> *I look at whoever asks me the question, and I tell them*
> *I wake up every day grateful for everything I have in my*
> *life. . . . I get to wake up every day and do a job I love. I get*

*to be challenged to push myself in many ways on many levels,
and I am always learning, always moving forwards. Not
many people can say that, and when I look at my life this
way, I consider myself very lucky. What do I have to be down
about? I have so much to be happy for.*

*There is no point in dwelling on what might or could have
been. The past has happened and cannot be changed; it can
only be accepted. Life is much simpler and much happier
when you always look at what you can do, not what you
can't.*

Every day is a good day.[5]

Henry's positive approach to life is a choice he has made,
and his mission to "always look at what you can do, not what
you can't" is an idea that we can all be helped by following.

Henry's approach of gratitude for what he has in life is
one that has been studied by psychologists and linked to
a range of positive outcomes. Psychologist Robert Emmons
specializes in gratitude and has found that it is critical to
people's well-being.[6] He learned that grateful people are
more happy, energetic, emotionally intelligent, and less
likely to be depressed, lonely, or anxious.[7] Importantly, he
shows that people are not more grateful because they are
happier, but that they can be trained to be more grateful
and happiness results from it. When researchers train people to become more grateful, those people become happier
and more optimistic.

Henry's choice to be grateful—for the "little things"—has
had a profound impact on his life and what he has been able
to achieve, even after being given a giant obstacle to overcome. Henry's life illustrates not only the impact of gratitude

but, more generally, the power of self-belief in achieving what many would see as impossible.

When we give up on something and decide we cannot do it, it is rarely because of actual limits;[8] instead, it is because we have decided we cannot do it. We are all susceptible to this negative and fixed thinking, but we become particularly susceptible to it when we age and start to feel that we are not as physically or mentally strong as we once were. Dayna Touron, at the University of North Carolina, has conducted research studies with adults over the age of sixty and shown that even aging is, in part, a result of our minds.[9] The adults in her study had to compare lists of words to see which words were the same in two lists. The adults could perform the task from memory (the researchers checked), but many did not trust their memories and instead laboriously cross-checked the lists, searching for similar words.

In another study of younger and older people working on calculations,[10] the researchers found that the younger people recalled previous answers and used them, but the older people chose to perform the calculations from scratch each time. As in the first experiment, the older people had developed memories they could use, but they did not trust them and so did not use them, which limited their effectiveness. The researchers also found that this memory avoidance and lack of confidence limited the adults' performance in their everyday activities. When we believe that we cannot do something, our limiting ideas often become self-fulfilling, and when we believe that we can do something, usually we can.

The knowledge that our brains are constantly changing and that we can learn complex skills at any age could help the scores of older adults who believe they are in decline and there is nothing they can do about it. Many people, as they age, start to believe they are capable of less, and this changes many decisions they make in their lives. Because they believe they can do less, they actually do less, which results in the cognitive decline they fear. Instead of retiring to a life of minimal activity, research tells us that we would be helped by filling our retirement years with new challenges and learning opportunities. Research has shown that elderly people who pursue more leisure activities have a 38 percent lower risk of developing dementia.[11]

Denise Park set out to study brain growth in older people and assigned groups of people to different activities for fifteen hours a week for three months.[12] The groups either learned skills such as quilting or photography, which required following detailed instructions and using long-term memory and attention, or engaged in more passive activities such as listening to classical music. At the end of the three months, it was only the people learning quilting or photography whose brains showed significant lasting changes to the medial frontal, lateral temporal, and parietal cortex—all areas associated with attention and concentration. A number of studies have pointed to the benefits of new hobbies that are unfamiliar and involve prolonged engagement for bringing about the greatest brain growth. Taking on a new hobby or class that involves concentration and struggle could give people substantial brain benefits that extend across their whole lives.

Some of the people we interviewed talked about a really important part of becoming limitless—when they changed their mindset, they realized they could do anything and that other people could no longer put obstacles in their way.

When they faced obstacles, they found a way around them, developing new strategies and new approaches to try. When we let go of limits and embrace the idea that anything can be achieved, it can change many aspects of our lives.

Beth Powell has been using her knowledge of brain growth, multidimensionality, and mindset to help the young people she works with map out new futures for themselves. She works at a school for students with special needs and has seen what she describes as "miracles" happen for the students when teachers and students believe in new possibilities and open pathways. But recently Beth found that she needed to apply her growth mindset to her own difficult medical situation.

Beth was forced to stop working when she developed serious medical problems and doctors could not find out what was wrong. The tests they conducted did not show any problems, so the doctors concluded that she must be fine. At this point many people would have given in and accepted what the doctors were saying. But Beth realized she could apply the approach she uses with her students to herself. Often students come to Beth with serious concerns, even though test scores do not reveal any particular disability or area of need. The students, however, know something is amiss. Beth's response is to consider each student as a whole person, not as a series of test scores, and take their concerns seriously. Beth applied this thinking to herself and decided to seek the help of holistic doctors. They were able to uncover the cause of Beth's illness, and she received the help she needed, which led to her recovery.

Beth recalled that she was at the point of going on disability pay and giving up work altogether when she remembered how students were transformed when people in their lives who knew about brain change worked with them. She realized that her situation was the same, and she needed to believe in change for her body and her medical situation, just as she believed in change for her students. This caused Beth to find a way around the roadblock she was facing—traditional doctors' inability to diagnose and treat her condition. The actions Beth took in those moments exemplify the approach of many of those we interviewed who became limitless. They faced roadblocks, but they had developed a new determination to find ways around them. They refused to take no for an answer, and they used different methods and took different approaches to the problems they faced, even when others were telling them to accept limiting judgments.

Now Beth is back at school helping students with special educational needs break free from the labels and limitations they have been diagnosed with. She reflects that she is even more committed to helping her students, because she experienced being written off herself. Beth said something to me in our interview that struck me as critically important. She told me that when students are referred to her school, she is often told that they "have such severe behavioral issues they cannot learn." But Beth refuses to believe this, knowing what might be possible for students if they are given access to learning and people who believe in them.

I am often asked by teachers what they can do with their "unmotivated" students. It is my firm belief that all students want to learn, and they only act unmotivated because someone, at some time in their lives, has given them the idea that they cannot be successful. Once students let go of these damaging ideas and someone opens a learning pathway for them, the lack of motivation goes away.

Beth's belief in students comes from many years of seeing incredible change. Because she teaches from the perspective of a growth mindset and makes use of multidimensional strategies, her students come to her school with learning differences but leave with a new future, free from negative labels and perceptions. She reflected, "Miracles are just normal to me now because of the brain science."

Embracing the knowledge that we can all change and grow and that limitations can be rejected is the first key to becoming unlocked and approaching life with a new limitless mind. This often allows people to let go of the idea that they are not good enough—and the importance of this particular change cannot be emphasized enough. Many people go through life feeling inadequate, often because a teacher, a boss, or, sadly, a parent or other family member has made them feel that way. When people feel they are not good enough, every failure or mistake is another opportunity to beat themselves up. When people realize that negative, limiting ideas are untrue, that any change can happen, and that times of struggle and failure are positive for brain growth, they stop feeling ashamed and start to feel empowered.

A second important key is knowing that times of struggle and making mistakes are good for our brains. There are two ways to approach mistakes that we make—negatively, with regret, or positively, with the idea that they will be an opportunity for learning, for brain growth, and for greater outcomes. I practice taking a positive approach to mistakes, thinking about the positive outcomes that may emerge from them on a daily basis. Sometimes mistakes are benign, and they can easily be fixed. At other times they may have real and negative consequences—initially—although later positive outcomes often emerge. Mistakes are a part of life, and the more courageous your life choices, the more mistakes you will probably make. Embracing mistakes will not make any difference in the number made, but you can choose to view them positively or negatively. If you choose the former, it will help you to become limitless.

Martin Samuels is a medical doctor who takes a positive approach to mistakes, despite being in a profession that works hard to outlaw mistakes. The medical world recognizes that mistakes can cost lives, and leaders publish papers and directives urging the avoidance of mistakes at all costs. This must challenge doctors who want to take a mistakes-friendly approach to life.

Samuels is unusual among doctors in embracing mistakes, recognizing that it is through mistakes that knowledge develops. Instead of beating himself up about them, he keeps a careful record of his mistakes, categorizes them, and shares them at conferences and other venues. In his blog post entitled "In Defense of Mistakes," he states that without errors "there would be no further evolution of medical thought" and if doctors accepted rather than feared mistakes and saw them as opportunities for learning rather than shame, they could actually focus on the real enemy, which is illness.[13] This open and positive approach toward mistakes has helped Samuels learn from them, become a better doctor, and help others in their path of learning and growth.

Embracing struggle—and choosing pathways that are hard—is just as important. If you settle into routines and do the same thing every day, it is unlikely that your brain will grow new pathways and connections. But if you constantly challenge yourself and embrace struggle, following new approaches and encountering new ideas, you will develop a sharpness that will enhance all aspects of your life.

Another important key to being limitless is approaching life through a multidimensional lens. This involves seeing the many different ways that problems—and life—can be approached. This can help the learning of content in any subject and at any level, from preschool to graduate school to everyday life. If you are stuck on an assignment or problem, it is highly likely that if you think about it differently and take a contrasting approach, moving, for example, from words to tables, numbers to visuals, or algorithms to graphs, solutions will emerge.

In 2016 a remarkable event took place. A mathematics problem that had never been solved, despite the many mathematicians who tried, was tackled by two young computer scientists.[14] The problem involved dividing a continuous object, whether a cake or a tract of land, equally, that is, so each party was satisfied with the piece received. Mathematicians had constructed a proof that worked but it was "unbounded," meaning it would need to run for a million or billion steps or any large number, depending on the preferences of the players. Some mathematicians had decided that the unbounded proof was the best they could ever find.

But the two young men decided to approach the problem differently. They did not have the wealth of knowledge that the mathematicians who tried it had, but that lack of knowledge helped them. They were not tied down by what they knew, as many people are; they were liberated by knowing less but being able to approach the problem with creativity.

Many people talked about the audacity of this event—the solving of a hard mathematics problem not by mathematicians, but by two young people without extensive mathematics knowledge. But knowledge can sometimes be inhibiting; it can stifle creative thinking[15] and lead people to use methods from a domain that they should look outside of. The two computer scientists who made the breakthrough believed that their success came from the fact that they had less knowledge than others, and this had allowed them to think differently.

Schools, educational establishments, and many companies do not, at this time, encourage different thinking enough, and sometimes nonroutine thinking is frowned upon or shut down entirely. Schools are designed to pass down established knowledge even when established knowledge is outdated, not the only way to think, and not the best way to solve problems. This is something that should change.

Before the problem I mentioned was solved, mathematicians believed that they just did not know enough to solve the problem. They effectively gave up on finding a solution that would work for any case and were planning to try to prove that the problem could not be solved. The computer scientists' fresh approach to the problem has now paved the way for new avenues of mathematical investigation.

In addition to thinking differently and creatively and embracing change, another key part of becoming limitless involves collaborating differently with people. One productive way to do this is to go into interactions with others with a willingness to share ideas even when you are not sure, instead of pretending to be an expert. When interactions are approached with an openness to learn and to expand, instead of a desire to look good, everybody is helped.

Open collaboration is much more likely to come about in companies and other institutions when it is modeled by managers and leaders. When they are the ones to say, "I don't know about this, but I would like to learn," others become emboldened to also embrace uncertainty and learning. When they are willing to listen and to expand their understanding, to be wrong and to say so openly, things change for those who work with them. When managers and leaders do this in their companies, and teachers and parents do this for students and children, then a culture of openness and growth is established.

Mark Cassar is the principal of a school in Toronto. He has worked to infuse mindset ideas throughout the K-8 school he runs. On my visit to the school, I was excited to see subjects being taught through a multidimensional approach. I sat and interviewed a range of young students, ages seven to ten, and was thrilled to hear the students talk about the school's approach, the mistake-friendly environment, and their positive self-beliefs—that they could learn anything.* The ideas I present in this book, about mindset, creativity, and multidimensionality, have influenced Mark in his work to change teaching as well as his work as a manager of people. In my interview with Mark, he talked about his revaluation of mistakes and the ways that had helped him as a manager:

> *I would say I'm far less critical of mistakes that I make,*
> *now. I'm actually a little bit easier on myself because I give*

---

* More details on Mark's school's approach and a video of his students can be seen here: https://www.youcubed.org/resources/an-example-of-a-growth-mindset-k-8-school/

myself a break to say, "Mark, mistakes are okay, so long as you learn from them." And I think as a principal I take a very similar approach with the kids. "It's okay. Mistakes are good so long as you learn from them. What did you learn from them? And how can we be better people moving forward?" I think it's changed me personally, but me as a professional too.

When you're dealing with people all day as I do, it's easy to misstep, and I think sometimes having a growth mindset allows you to be reflective and to say, "Okay, you know what? How did I manage this and can I do something different next time?" I think I would not have been able to do that as well prior to seeing all of your work, but now it's helped me to be much better at being reflective and a critical thinker, where I wouldn't have been before.

Well, I do remember a time where I was dealing with a student who made a pretty grand mistake in how they were conducting themselves in school. And I was rather judgmental in how I approached them, thinking I was right and they were wrong, only to find out later, when the facts became more clear, that they were actually right and I was wrong. And I realized, "Okay, you know what? Just because I'm the principal, I can't presume I have no accountability to myself to improve. Because even though you're the boss, so to speak, it doesn't mean you're beyond reproach, right?" Actually I went back to the student and said, "You know what? I think I made a mistake. You were right, I was wrong, and next time around I'll handle it a little bit differently." I think just in the day-to-day operations of how you conduct yourself, it's made a big difference for me.

Mark has developed a school in which every teacher has embraced a mindset and multidimensionality approach, and they have seen students' love of learning and their achievement increase because of it. One important part of the change they have worked on together is transforming assessment and testing. The teachers realized something important—that it is hard to tell students that mistakes are really useful for learning, but then penalize them on tests for every mistake they make.

The teachers still assess students, but instead of giving an unhelpful number and penalizing mistakes, they give what I often describe as the greatest gift teachers can give students—diagnostic comments on ways to improve based on a leveled rubric (an evaluative scoring guide). Mark said that at first the students looked for a number and that was all they cared about, which is often the result in a performance culture (rather than a learning culture). But now students see the rubric, they understand where they are in their learning, and they read the teacher's comments to know how to improve. This change in assessment is the ultimate way to share with students the message that growth and learning are what you value and that you can help them improve with guidance.[16] One of the rubrics from Mark's school can be found in Appendix II and more detail is available on youcubed.org.

Mark and his teaching staff are working to give students a limitless approach to life. This, sadly, contrasts with the perspectives of so many students and adults who have learned to view themselves and their situations in a fixed light. Some people have been locked into a negative way of thinking by their parents, who have made them feel that they are not good enough. Children can also become locked up because of classroom interactions and people who do not believe in them and assume they cannot learn. But they are also locked up when content is monotonous, boring, and repetitive and they cannot see a way to learn. There are myriad ways that the world we live in attempts to limit our belief in ourselves and our potential. And now, we have a better understanding of the keys necessary to tackle these obstacles, regardless of the situation we find ourselves in.

A change from believing there are limits to learning, and life, to believing that anything can be learned or achieved is a change from a fixed to a growth mindset. When we make this change, it has a transformative effect on our lives. We stop thinking we are not good enough and start to take more risks. When we add in the knowledge that struggle and failure are important for our brains and can be thought of as opportunities for learning, greater liberation is possible. We start to see our minds as fluid instead of fixed and start to see infinite life possibilities. When we also learn that we can take a multidimensional approach to academic content and to life problems and that we can collaborate with others as partners instead of competitors, it changes not only our thinking about potential, but also all the interactions that permeate our lives. We realize that obstacles cannot be put in our way, that we can always develop strategies to overcome them.

We change our minds, but also our hearts and spirits, as we became more flexible, fluid, and adaptable. If we face roadblocks, we find ways around them, refusing to accept the negative judgments of others. Some of us will change not only our own lives; as we start to see ourselves as leaders and ambassadors, we can help others live limitless lives as well. Even young children who learn about brain growth and change, mistakes and multidimensionality, often take it upon themselves to share the news with all the people around them.

Shawn Achor's book *The Happiness Advantage* sets out to dispel an important limiting myth. Many people believe that they will become happier if they work harder, get a better job, find a perfect partner, lose ten pounds, and so on (substitute any goal of your own). But a range of research studies have shown that this is backward thinking and that when people become positive, they become more motivated, engaged, creative, and productive in all sorts of ways. As he says: "Happiness fuels success, not the other way around."[17] Achor illustrates the importance of positive thinking with a particularly adorable, yet powerful personal story from his childhood that I want to share with you.

Shawn was seven at the time, playing with his five-year-old sister on the top of their bunk beds. As the older sibling, he had been told that he was responsible for the two of them as they played quietly while his parents took a nap. As the older brother, he also got to decide what they played, so he suggested a military battle between his G.I. Joes and her unicorns and My Little Ponies!

The two siblings lined up their toys, but then in a moment of excitement, the little girl fell off the bunk bed. Shawn heard a crash and peered over to see that Amy had landed on her hands and knees, on all fours. In that moment he was worried, not only that she might be hurt, but because he could see she was about to erupt in a loud wail that would wake his parents. Then, as he recalls:

Crisis is the mother of all invention, so I did the only thing
my frantic little seven-year-old brain could think to do.
I said, "Amy, wait! Wait. Did you see how you landed?
No human lands on all fours like that. You . . . you're a
unicorn!"[18]

He knew that there was nothing in the world his sister wanted more than to be a unicorn, and in that moment she chose not to wail but instead to be excited about her "new-found identity as a unicorn"! A smile broke out on her face, and she climbed back up on the bed to continue playing.

To me this is a powerful story, because it is about the kinds of moments of choice that fill our lives. We can choose to be negative or positive, and what we choose changes our outlook and our future. We do not always have an older brother to give us the idea we can be a unicorn; instead, we have knowledge—of ways to handle failure, to develop a positive mindset, to use multidimensionality and creativity in solving problems—and most important, the knowledge that how we respond will shape future outcomes. A change in mindset not only changes the way we think about reality; it changes our reality.

As an educator of many years, I have met students—children and adults—who have been held back by limits. Fortunately, my work also allows me to meet children and young adults who have learned that they can do anything and that nothing will limit them, and I watch the ways that their positive thinking impacts everything around them. When they fall from the metaphorical bunk bed, which we all do from time to time, they do not cry; instead, they decide, "I am a unicorn!"

So my final advice for you is to embrace struggle and failure, to take risks, and to not let people obstruct your pathways. If a barrier or roadblock is put in your way, find a way around it, take a different approach. If you are in a job and you would rather take on something someone else has always done, explore that new avenue. If your job does not allow you to work outside of limits, then maybe you should look for another job. Do not accept a life of limits. Instead of looking back on things that have gone badly, look forward and be positive about opportunities for learning and improvement. See others as collaborators, with whom you can grow and learn. Share uncertainty with them and be open to different ways of thinking. If you are an educator or manager, find out how your students or colleagues think. Value multiple ways of thinking, seeing, and working. The most beautiful part of problem solving is multidimensionality, the multiple ways any problem can be seen or solved. This is the diversity of life that is so important to embrace and value—in mathematics, in art, in history, in management, in sports, in anything.

Try living a single day of your life with a limitless approach and you will notice the difference. Unlock the pathways of others and know that you will be changing their lives for the better, and they can go on and change the lives of others. There may be nothing more important for our own or for our learners' lives than knowing that we can always reach for the stars. Sometimes we won't succeed, and that is okay, but we will always be helped by setting out on the journey—especially if the perspective we take on that journey is truly limitless.

ACKNOWLEDGMENTS

I am deeply grateful to all the people I interviewed for this book—teachers, leaders, parents, writers, and others. They opened their hearts and shared their stories. In doing so they were vulnerable—they told me about how life was before they learned about these ideas. Many of the people shared that they tried to be "perfect" and were afraid of not knowing—they had been told by others they were not a math person or another type of person, and they were discouraged from learning to high levels. They also shared their journey of change, and in many cases the ways they were now inspiring others. Not all the people I interviewed made it into the book, as space was short—but I am deeply grateful to them all:

Cherry Agapito

Caleb Austin

Terese Barham

Sara Boone

Angela Brennan

Jennifer Brich

Jim Brown

Heather Buske

Jodi Campinelli

Mark Cassar

Evelyn Chan

Holly Compton

Kate Cook

Stephanie Diehl

Robin Dubiel

Margriet Faber

Kirstie Fitzgerald

Shelley Fritz

Mariève Gagnè

Marta Garcia

Karen Gauthier

Allison Giacomini

Rene Grimes

Margaret Hall

Judith Harris

Suzanne Harris

Leah Haworth

Meg Hayes

Catherine Head

Susan Jachymiak

Lauren Johnson

Theresa Lambert

Linda Lapere

Zandi Lawrence

Lucia MacKenzie

Jean Maddox

Sunil Reddy Mayreddy

Chelsea McClellan

Sara McGee

Shana McKay

Adele McKew

Jesse Melgares

Gail Metcalf

Crystal Morey

Jenny Morrill

Pete Noble

Marc Petrie

Meryl Polak

Beth Powell

Justin Purvis

Nancy Qushair

Sunil Reddy

Evette Reece

Kate Rizzi

Daniel Rocha

Tami Sanders

Jennifer Schaefer

Michelle Scott

Erica Sharma

Nina Sudnick

Angela Thompson

Carrie Tomc

Laura Wagenman

Ben Woodford

I am always extremely grateful to my family for any book I write, as they have to put up with me not being there for them. I have two amazing daughters, Ariane and Jaime, and they light up my life every day.

I am also grateful to my youcubed cofounder and good friend Cathy Williams—she is always my thought partner, in some cases she draws pictures for me, and often she puts up with my most crazy thoughts—and encourages them. Viva La Revolution, Cathy!

Also important is the dynamic team at youcubed; I could

not have written this book without them. They helped me with interviews, and with their ongoing support for all of our work. They are: Montserrat Cordero, Suzanne Corkins, Kristina Dance, Jack Dieckmann, Jessica Method, and Estelle Woodbury. My doctoral students Tanya LaMar and Robin Anderson also gave invaluable support and help.

Beyond the many teachers I interviewed for this book, I am inspired by teachers on a daily basis. There may be some who give fixed ideas to children, but there are many others who believe in all students, who spend countless hours preparing engaging lessons for students and who go above and beyond what anyone should reasonably be asked to do for their job. If we gave more decisions to teachers about what and how students should learn, we would be in a much better place than we are now. Thank you to all the teachers who I have had the honor of talking to and learning from over the last few years.

RESOURCES
to Help Change Mindsets
and Approaches

Four Boosting Messages for Students:
    **https://www.youcubed.org/resources/four-boosting
-messages-jo-students/**

Free Online Class, in English and Spanish, to Improve
Students' Mindset and Approach to Mathematics:
    https://www.youcubed.org/online-student-course/

A Range of Mindset Videos for Students:
    **https://www.youcubed.org/resource/mindset-boosting
-videos/**

Rethinking Giftedness Film:
    https://www.youcubed.org/rethinking-giftedness-film/

Different Experiences with Maths Facts Film:
    **https://www.youcubed.org/resources/different
-experiences-with-math-facts/**

Visual Creative Maths Tasks:
    https://www.youcubed.org/tasks/

Free Downloadable Posters:
> https://www.youcubed.org/resource/posters/

Two Online Courses for Mathematics Teachers and Parents:
> https://www.youcubed.org/online-teacher-courses/

K–8 Book Series:
> https://www.youcubed.org/resource/k-8-curriculum/

A Range of Short, Readable News Articles on the Book's Ideas:
> https://www.youcubed.org/resource/in-the-news/

APPENDIX I

Examples of Numerical and Visual Approaches to Math Problems

Here are two standard math questions with visual solutions. These are the types of questions that may have produced anxiety and math hatred in school, with good reason. I have written at length about the damage of fake word problems and contexts that students are meant to partly believe, while ignoring everything they know about the real situation. But please have a look at the different ways of solving them, as an illustration of what is possible when we think visually.

This problem is adapted from one used by a wonderful mathematics educator, Ruth Parker. She poses this question:

A man wants to buy ¼ of a pound of turkey. He goes into a shop that gives him 3 slices that weigh ⅓ of a pound. What proportion of the 3 slices does he need?

A numerical approach:	A visual approach:
3 slices = ⅓ pound	○○○ = ⅓ pound
x slices = ¼ pound	
	○○○
⅓ x = ¾	○○○ = ⅓ pound
	○○○
x = ¾	
	○⦶○
	⊖⦶⊖ = ¼ pound
	○⦶○ (or 2 ¼ slices)

The second is one of those awful, unrealistic word problems that fill mathematics textbooks:

Jo and Tesha each have a number of cards in the ratio 2:3. Tesha and Holly have a number of cards in the ratio 2:1. If Tesha has 4 more cards than Jo, how many cards does Holly have? Give an answer and briefly explain your reasoning.

A numerical approach:

Jo and Tesha 2:3

Tesha and Holly 2:1

Jo and Tesha's cards are divided into 5, with a 2:3 ratio.

Tesha has ⅕ more than Jo. Tesha has 4 more cards.

⅕ = 4

1 = 20

So together they have 20 cards.

Jo has ⅖ × 20 and Tesha has ⅗ × 20.

Jo has 8 and Tesha has 12.

Tesha and Holly are 2:1, so Holly has 6 cards.

A visual approach:

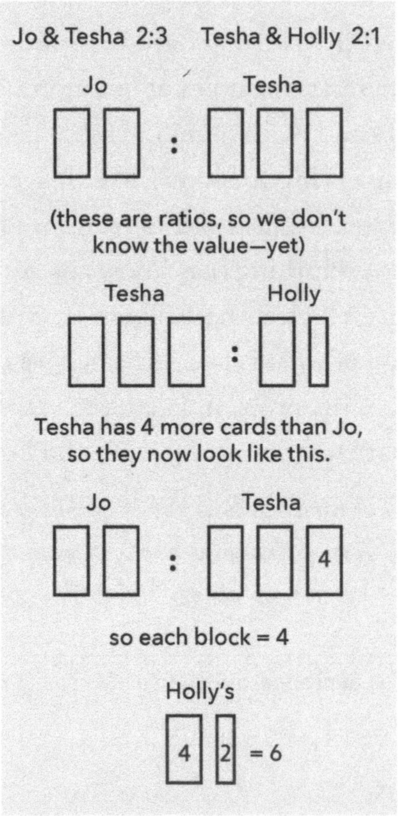

Jo & Tesha 2:3   Tesha & Holly 2:1

Jo   Tesha

☐ ☐   :   ☐ ☐ ☐

(these are ratios, so we don't know the value—yet)

Tesha   Holly

☐ ☐ ☐   :   ☐ ☐

Tesha has 4 more cards than Jo, so they now look like this.

Jo   Tesha

☐ ☐   :   ☐ ☐ ☐4

so each block = 4

Holly's

☐4 ☐2 = 6

APPENDIX II

A Sample Rubric

Here is a rubric from Mark Cassar's school. In this rubric the teacher decides whether a student has met the area of learning described in the "criteria" and includes feedback for the student on ways to improve. In this case the rubric also reflects a conversation the teacher had with the student to clarify understanding.

Toothpicks Problem (Patterning)
Assessment For/As Learning

Criteria	1	2	3	4	Feedback
Create, identify, extend patterns		✓			*"How could you find out the number of toothpicks for the 6th term?"*
Make a table of values for a pattern	✓				*A table of values (t-chart) will help you determine the pattern rule*
Communicate math thinking in writing and pictures (communication and representation) *& oral*			✓		*Conversation with student "tell me about"* 22 $\overline{6}$

1 = expectation not met; **2** = approaching expectation;
3 = meets expectation; **4** = exceeds expectation

* *I had a conference with the student; her changes are noted on #2b.*

"I am adding together to find the total amount."

NOTES

Introduction

1 Sue Johnston-Wilder, Janine Brindley, and Philip Dent, *A Survey of Mathematics Anxiety and Mathematical Resilience Among Existing Apprentices* (London: Gatsby Charitable Foundation, 2014).

2 Sara Draznin, "Math Anxiety in Fundamentals of Algebra Students," *The Eagle Feather*, Honors College, Univ. of North Texas, January 1, 1970, http://eaglefeather.honors.unt.edu/2008/article/179#.W-idJS2ZNMM; N. Betz, "Prevalence, Distribution, and Correlates of Math Anxiety in College Students," *Journal of Counseling Psychology* 25/5 (1978): 441–48.

3 C. B. Young, S. S. Wu, and V. Menon, "The Neurodevelopmental Basis of Math Anxiety," *Psychological Science* 23/5 (2012): 492–501.

4 Daniel Coyle, *The Talent Code: Greatness Isn't Born. It's Grown. Here's How.* (New York: Bantam, 2009).

5 Michael Merzenich, *Soft-Wired: How the New Science of Brain Plasticity Can Change Your Life* (San Francisco: Parnassus, 2013).

6 Merzenich, *Soft-Wired*.

7 Anders Ericsson and Robert Pool, *Peak: Secrets from the New Science of Expertise* (New York: Houghton Mifflin Harcourt, 2016).

8 Ericsson and Pool, *Peak*, 21.

9 Carol S. Dweck, *Mindset: The New Psychology of Success* (New York: Ballantine, 2006).

10 Carol S. Dweck, "Is Math a Gift? Beliefs That Put Females at Risk," in Stephen J. Ceci and Wendy M. Williams, eds., *Why Aren't More Women in Science? Top Researchers Debate the Evidence* (Washington, DC: American Psychological Association, 2006).

11 D. S. Yeager et al., "Breaking the Cycle of Mistrust: Wise Interventions to Provide Critical Feedback Across the Racial Divide," *Journal of Experimental Psychology: General* 143/2 (2014): 804.

Chapter 1

1 Michael Merzenich, *Soft-Wired: How the New Science of Brain Plasticity Can Change Your Life* (San Francisco: Parnassus, 2013), 2.

2 Norman Doidge, *The Brain That Changes Itself* (New York: Penguin, 2007).

3 Doidge, *The Brain That Changes Itself*, 55.

4 E. Maguire, K. Woollett, and H. Spiers, "London Taxi Drivers and Bus Drivers: A Structural MRI and Neuropsychological Analysis," *Hippocampus* 16/12 (2006): 1091–101.

5 K. Woollett and E. A. Maguire, "Acquiring 'The Knowledge' of London's Layout Drives Structural Brain Changes," *Current Biology* 21/24 (2011): 2109–14.

6 Elise McPherson et al., "Rasmussen's Syndrome and Hemispherectomy:

Girl Living with Half Her Brain," *Neuroscience Fundamentals*, http://www
.whatsonxiamen.com/news11183.html.

7   Doidge, *The Brain That Changes Itself*, xix.

8   Doidge, *The Brain That Changes Itself*, xx.

9   A. Dixon, editorial, *FORUM* 44/1 (2002): 1.

10 Sarah D. Sparks, "Are Classroom Reading Groups the Best Way to Teach
    Reading? Maybe Not," *Education Week*, August 26, 2018, http://www.edweek
    .org/ew/articles/2018/08/29/are-classroom-reading-groups-the-best-way.html.

11 Sparks, "Are Classroom Reading Groups the Best Way to Teach Reading?
    Maybe Not."

12 Jo Boaler, *Mathematical Mindsets: Unleashing Students' Potential Through Creative
    Math, Inspiring Messages and Innovative Teaching* (San Francisco: Jossey-Bass,
    2016).

13 Jo Boaler et al., "How One City Got Math Right," *The Hechinger Report*, Octo-
    ber 2018, https://hechingerreport.org/opinion-how-one-city-got-math-right/.

14 Lois Letchford, *Reversed: A Memoir* (Irvine, CA: Acorn, 2018).

15 Doidge, *The Brain That Changes Itself*, 34.

16 K. Lewis and D. Lynn, "Against the Odds: Insights from a Statistician with
    Dyscalculia," *Education Sciences* 8/2 (2018): 63.

17 T. Iuculano et al., "Cognitive Tutoring Induces Widespread Neuroplasticity
    and Remediates Brain Function in Children with Mathematical Learning
    Disabilities," *Nature Communications* 6 (2015): 8453, https://doi.org/10.1038
    /ncomms9453.

18 Sarah-Jane Leslie, et al., "Expectations of Brilliance Underlie Gender Distri-
    butions Across Academic Disciplines," *Science* 347/6219 (2015): 262–65.

19 Seth Stephens-Davidowitz, "Google, Tell Me: Is My Son a Genius?" *New York
    Times*, January 18, 2014, https://www.nytimes.com/2014/01/19/opinion
    /sunday/google-tell-me-is-my-son-a-genius.html.

20 D. Storage et al., "The Frequency of 'Brilliant' and 'Genius' in Teaching
    Evaluations Predicts the Representation of Women and African Americans
    Across Fields," *PLoS ONE* 11/3 (2016): e0150194, https://doi.org/10.1371
    /journal.pone.0150194.

21 Piper Harron, "Welcome to Office Hours," *The Liberated Mathematician*, 2015,
    http://www.theliberatedmathematician.com.

22 Eugenia Sapir, "Maryam Mirzakhani as Thesis Advisor," *Notices of the AMS*
    65/10 (November 2018): 1229–30.

23 At the time of this writing, the film, which can be seen at http://www
    .youcubed.org/rethinking-giftedness-film, has had 62,000 views.

24 Daniel Coyle, *The Talent Code: Greatness Isn't Born. It's Grown. Here's How.*
    (New York: Bantam, 2009), 178.

25 Anders Ericsson and Robert Pool, *Peak: Secrets from the New Science of
    Expertise* (New York: Houghton Mifflin Harcourt, 2016).

Chapter 2

1   J. S. Moser et al., "Mind Your Errors: Evidence for a Neural Mechanism
    Linking Growth Mind-set to Adaptive Posterror Adjustments," *Psychological
    Science* 22/12 (2011): 1484–89.

2   Daniel Coyle, *The Talent Code: Greatness Isn't Born. It's Grown. Here's How.*
    (New York: Bantam, 2009).

3   J. A. Mangels, et al., "Why Do Beliefs About Intelligence Influence Learning
    Success? A Social Cognitive Neuroscience Model," *Social Cognitive and
    Affective Neuroscience* 1/2 (2006): 75–86, http://academic.oup.com/scan
    /article/1/2/75/2362769.

4   Moser et al., "Mind Your Errors."
5   Coyle, *The Talent Code*, 2–3.
6   Coyle, *The Talent Code*, 3–4.
7   Coyle, *The Talent Code*, 5.
8   Moser et al., "Mind Your Errors."
9   Anders Ericsson and Robert Pool, *Peak: Secrets from the New Science of Expertise* (New York: Houghton Mifflin Harcourt, 2016), 75.
10 James W. Stigler and James Hiebert, *The Teaching Gap: Best Ideas from the World's Teachers for Improving Education in the Classroom* (New York: Free Press, 1999).
11 Elizabeth Ligon Bjork and Robert Bjork, "Making Things Hard on Yourself, but in a Good Way: Creating Desirable Difficulties to Enhance Learning," in Morton Ann Gernsbacher and James R. Pomeratz, eds., *Psychology and the Real World* (New York: Worth, 2009), 55–64, https://bjorklab.psych.ucla.edu/wp-content/uploads/sites/13/2016/04/EBjork_RBjork_2011.pdf.
12 J. Boaler, K. Dance, and E. Woodbury, "From Performance to Learning: Assessing to Encourage Growth Mindsets," *youcubed*, 2018, tinyurl.com/A4Lyoucubed.
13 Coyle, *The Talent Code*, 5.

Chapter 3

1   O. H. Zahrt and A. J. Crum, "Perceived Physical Activity and Mortality: Evidence from Three Nationally Representative U.S. Samples," *Health Psychology* 36/11 (2017): 1017–25, http://dx.doi.org/10.1037/hea0000531.
2   B. R. Levy et al., "Longevity Increased by Positive Self-Perceptions of Aging," *Journal of Personality and Social Psychology* 83/2 (2002): 261–70, https://doi.org/10.1037/0022-3514.83.2.261.
3   B. R. Levy et al., "Age Stereotypes Held Earlier in Life Predict Cardiovascular Events in Later Life," *Psychological Science* 20/3 (2009): 296–98, https://doi.org/10.1111/j.1467-9280.2009.02298.x.
4   Levy et al., "Age Stereotypes Held Earlier in Life."
5   A. J. Crum and E. J. Langer, "Mind-Set Matters: Exercise and the Placebo Effect," *Psychological Science* 18/2 (2007): 165–71, https://doi.org/10.1111/j.1467-9280.2007.01867.x.
6   V. K. Ranganathan et al., "From Mental Power to Muscle Power—Gaining Strength by Using the Mind," *Neuropsychologia* 42/7 (2004): 944–56.
7   N. F. Bernardi et al., "Mental Practice Promotes Motor Anticipation: Evidence from Skilled Music Performance," *Frontiers in Human Neuroscience* 7 (2013): 451, https://doi.org/10.3389/fnhum.2013.00451.
8   K. M. Davidson-Kelly, "Mental Imagery Rehearsal Strategies for Expert Pianists," *Edinburgh Research Archive*, November 26, 2014, https://www.era.lib.ed.ac.uk/handle/1842/14215.
9   D. S. Yeager, K. H. Trzesniewski, and C. S. Dweck, "An Implicit Theories of Personality Intervention Reduces Adolescent Aggression in Response to Victimization and Exclusion," *Child Development* 84/3 (2013): 970–88.
10 P. B. Carr, C. S. Dweck, and K. Pauker, "'Prejudiced' Behavior Without Prejudice? Beliefs About the Malleability of Prejudice Affect Interracial Interactions," *Journal of Personality and Social Psychology* 103/3 (2012): 452.
11 L. S. Blackwell, K. H. Trzesniewski, and C. S. Dweck, "Implicit Theories of Intelligence Predict Achievement Across an Adolescent Transition: A Longitudinal Study and an Intervention," *Child Development* 78/1 (2007): 246–63.

12 J. S. Moser et al., "Mind Your Errors: Evidence for a Neural Mechanism Linking Growth Mind-set to Adaptive Posterror Adjustments," *Psychological Science* 22/12 (2011): 1484–89.

13 E. A. Gunderson et al., "Parent Praise to 1- to 3-Year-Olds Predicts Children's Motivational Frameworks 5 Years Later," *Child Development* 84/5 (2013): 1526–41.

14 Carol S. Dweck, "The Secret to Raising Smart Kids," *Scientific American Mind* 18/6 (2007): 36–43, https://doi.org/10.1038/scientificamericanmind 1207-36.

15 Carol S. Dweck, "Is Math a Gift? Beliefs That Put Females at Risk," in Stephen J. Ceci and Wendy M. Williams, eds., *Why Aren't More Women in Science? Top Researchers Debate the Evidence* (Washington, DC: American Psychological Association, 2006).

16 Blackwell, Trzesniewski, and Dweck, "Implicit Theories of Intelligence Predict Achievement."

17 Angela Duckworth, *Grit: The Power of Passion and Perseverance* (New York: Scribner, 2016).

18 J. Boaler, K. Dance, and E. Woodbury, "From Performance to Learning: Assessing to Encourage Growth Mindsets," *youcubed*, 2018, tinyurl.com /A4Lyoucubed.

19 H. Y. Lee et al., "An Entity Theory of Intelligence Predicts Higher Cortisol Levels When High School Grades Are Declining," *Child Development*, July 10, 2018, https://doi.org/10.1111/cdev.13116.

20 Anders Ericsson and Robert Pool, *Peak: Secrets from the New Science of Expertise* (New York: Houghton Mifflin Harcourt, 2016).

21 Carol S. Dweck, *Mindset: The New Psychology of Success* (New York: Ballantine, 2006), 257.

22 Christine Gross-Loh, "How Praise Became a Consolation Prize," *The Atlantic*, December 16, 2016.

Chapter 4

1 Alfie Kohn, "The 'Mindset' Mindset," *Alfie Kohn*, June 8, 2018, http://www .alfiekohn.org/article/mindset/.

2 V. Menon, "Salience Network," in Arthur W. Toga, ed., *Brain Mapping: An Encyclopedic Reference*, vol. 2 (London: Academic, 2015), 597–611.

3 J. Park and E. M. Brannon, "Training the Approximate Number System Improves Math Proficiency," *Psychological Science* 24/10 (2013): 2013–19, https://doi.org/10.1177/0956797613482944.

4 I. Berteletti and J. R. Booth, "Perceiving Fingers in Single-Digit Arithmetic Problems," *Frontiers in Psychology* 6 (2015): 226, https://doi.org/10.3389 /fpsyg.2015.00226.

5 M. Penner-Wilger and M. L. Anderson, "The Relation Between Finger Gnosis and Mathematical Ability: Why Redeployment of Neural Circuits Best Explains the Finding," *Frontiers in Psychology* 4 (2013): 877, https://doi .org/10.3389/fpsyg.2013.00877.

6 M. Penner-Wilger et al., "Subitizing, Finger Gnosis, and the Representation of Number," *Proceedings of the 31st Annual Cognitive Science Society* 31 (2009): 520–25.

7 S. Beilock, *How the Body Knows Its Mind: The Surprising Power of the Physical Environment to Influence How You Think and Feel* (New York: Simon and Schuster, 2015).

8 Anders Ericsson and Robert Pool, *Peak: Secrets from the New Science of Expertise* (New York: Houghton Mifflin Harcourt, 2016).

9   A. Sakakibara, "A Longitudinal Study of the Process of Acquiring Absolute Pitch: A Practical Report of Training with the 'Chord Identification Method,'" *Psychology of Music* 42/1 (2014): 86–111, https://doi.org/10.1177/0305735612463948.

10 Thomas G. West, *Thinking Like Einstein: Returning to Our Visual Roots with the Emerging Revolution in Computer Information Visualization* (New York: Prometheus Books, 2004).

11 Claudia Kalb, "What Makes a Genius?" *National Geographic*, May 2017.

12 Kalb, "What Makes a Genius?"

13 M. A. Ferguson, J. S. Anderson, and R. N. Spreng, "Fluid and Flexible Minds: Intelligence Reflects Synchrony in the Brain's Intrinsic Network Architecture," *Network Neuroscience* 1/2 (2017): 192–207.

14 M. Galloway, J. Conner, and D. Pope, "Nonacademic Effects of Homework in Privileged, High-Performing High Schools," *Journal of Experimental Education* 81/4 (2013): 490–510.

15 M. E. Libertus, L. Feigenson, and J. Halberda, "Preschool Acuity of the Approximate Number System Correlates with School Math Ability," *Developmental Science* 14/6 (2011): 1292–1300.

16 R. Anderson, J. Boaler, and J. Dieckmann, "Achieving Elusive Teacher Change Through Challenging Myths About Learning: A Blended Approach," *Education Sciences* 8/3 (2018): 98.

17 Anderson, Boaler, and Dieckmann, "Achieving Elusive Teacher Change."

18 J. Boaler, K. Dance, and E. Woodbury, "From Performance to Learning: Assessing to Encourage Growth Mindsets," *youcubed*, 2018, tinyurl.com/A4Lyoucubed.

Chapter 5

1   Claudia Kalb, "What Makes a Genius?" *National Geographic*, May 2017.

2   Sian Beilock, *Choke: What the Secrets of the Brain Reveal About Getting It Right When You Have To* (New York: Simon and Schuster, 2010).

3   A paper that gives advice on different ways to teach math facts conceptually and well—without fear or anxiety—is Jo Boaler, Cathy Williams, and Amanda Confer, "Fluency Without Fear: Research Evidence on the Best Ways to Learn Math Facts," *youcubed*, January 28, 2015, https://www.youcubed.org/evidence/fluency-without-fear.

4   E. A. Maloney et al., "Intergenerational Effects of Parents' Math Anxiety on Children's Math Achievement and Anxiety," *Psychological Science* 26/9 (2015): 1480–88, https://doi.org/10.1177/0956797615592630.

5   S. L. Beilock et al., "Female Teachers' Math Anxiety Affects Girls' Math Achievement," *Proceedings of the National Academy of Sciences* 107/5 (2010): 1860–63.

6   Laurent Schwartz, *A Mathematician Grappling with His Century* (Basel: Birkhäuser, 2001).

7   Kenza Bryan, "Trailblazing Maths Genius Who Was First Woman to Win Fields Medal Dies Aged 40," *Independent*, July 15, 2017, https://www.independent.co.uk/news/world/maryam-mirzakhani-fields-medal-mathematics-dies-forty-iran-rouhani-a7842971.html.

8   Schwartz, *A Mathematician Grappling with His Century*, 30–31.

9   Norman Doidge, *The Brain That Changes Itself* (New York: Penguin, 2007), 199.

10 Doidge, *The Brain That Changes Itself*, 199.

11 K. Supekar et al., "Neural Predictors of Individual Differences in Response to Math Tutoring in Primary-Grade School Children," *PNAS* 110/20 (2013): 8230–35.

12 E. M. Gray and D. O. Tall, "Duality, Ambiguity, and Flexibility: A 'Proceptual' View of Simple Arithmetic," *Journal for Research in Mathematics Education* 25/2 (1994): 116–40.

13 W. P. Thurston, "Mathematical Education," *Notices of the American Mathematical Society* 37 (1990): 844–50.

14 Gray and Tall, "Duality, Ambiguity, and Flexibility."

15 Jo Boaler and Pablo Zoida, "Why Math Education in the U.S. Doesn't Add Up," *Scientific American*, November 1, 2016, https://www.scientificamerican .com/article/why-math-education-in-the-u-s-doesn-t-add-up.

16 Adam Grant, *Originals: How Non-Conformists Move the World* (New York: Penguin, 2016).

17 Grant, *Originals*, 9–10.

Chapter 6

1 U. Treisman, "Studying Students Studying Calculus: A Look at the Lives of Minority Mathematics Students in College," *College Mathematics Journal* 23/5 (1992): 362–72 (368).

2 Treisman, "Studying Students Studying Calculus," 368.

3 Organisation for Economic Co-operation and Development, *The ABC of Gender Equality in Education: Aptitude, Behaviour, Confidence* (Paris: PISA, OECD Publishing, 2015), https://www.oecd.org/pisa/keyfindings/pisa -2012-results-gender-eng.pdf.

4 OECD, *The ABC of Gender Equality in Education*.

5 M. I. Núñez-Peña, M. Suárez-Pellicioni, and R. Bono, "Gender Differences in Test Anxiety and Their Impact on Higher Education Students' Academic Achievement," *Procedia - Social and Behavioral Sciences* 228 (2016): 154–60.

6 Organisation for Economic Co-operation and Development, *PISA 2015 Results (Volume V): Collaborative Problem Solving* (Paris: PISA, OECD Publishing, 2017), https://doi.org/10.1787/9789264285521-en.

7 J. Decety et al., "The Neural Bases of Cooperation and Competition: An fMRI Investigation," *Neuroimage* 23/2 (2004): 744–51.

8 V. Goertzel et al., *Cradles of Eminence: Childhoods of More than 700 Famous Men and Women* (Gifted Psychology Press: 2004), 133–55.

9 Meg Jay, "The Secrets of Resilience," *Wall Street Journal*, November 10, 2017, https://www.wsj.com/articles/the-secrets-of-resilience-1510329202.

10 Jo Boaler, "Open and Closed Mathematics: Student Experiences and Understandings," *Journal for Research in Mathematics Education* 29/1 (1998): 41–62.

11 Jo Boaler, *Experiencing School Mathematics: Traditional and Reform Approaches to Teaching and Their Impact on Student Learning* (New York: Routledge, 2002).

12 J. Boaler and S. Selling, "Psychological Imprisonment or Intellectual Freedom? A Longitudinal Study of Contrasting School Mathematics Approaches and Their Impact on Adults' Lives," *Journal of Research in Mathematics Education* 48/1 (2017): 78–105.

13 J. Boaler and M. Staples, "Creating Mathematical Futures Through an Equitable Teaching Approach: The Case of Railside School," *Teachers' College Record* 110/3 (2008): 608–45.

14 Jo Boaler, "When Academic Disagreement Becomes Harassment and Persecution," October 2012, http://web.stanford.edu/~joboaler.

15 Shane Feldman, "Pain to Purpose: How Freshman Year Changed My Life," https://www.youtube.com/watch?v=BpMq7Q54cwI.

16 Jo Boaler, "Promoting 'Relational Equity' and High Mathematics

Achievement Through an Innovative Mixed Ability Approach," *British Educational Research Journal* 34/2 (2008): 167–94.

17 John J. Cogan and Ray Derricott, *Citizenship for the 21st Century: An International Perspective on Education* (London: Kogan Page, 1988), 29; Gita Steiner-Khamsi, Judith Torney-Purta, and John Schwille, eds., *New Paradigms and Recurring Paradoxes in Education for Citizenship: An International Comparison* (Bingley, UK: Emerald Group, 2002).

18 Boaler and Staples, "Creating Mathematical Futures."

19 Jenny Morrill and Paula Youmell, *Weaving Healing Wisdom* (New York: Lexingford, 2017).

Conclusion

1 Etienne Wenger, *Communities of Practice: Learning, Meaning, and Identity* (Cambridge: Cambridge Univ. Press, 1999).

2 Angela Duckworth, *Grit: The Power of Passion and Perseverance* (New York: Scribner, 2016).

3 Nicole M. Joseph, personal communication, 2019.

4 Henry Fraser, *The Little Big Things* (London: Seven Dials, 2018).

5 Fraser, *The Little Big Things*, 158–59.

6 R. A. Emmons and M. E. McCullough, "Counting Blessings Versus Burdens: An Experimental Investigation of Gratitude and Subjective Well-Being in Daily Life," *Journal of Personality and Social Psychology* 84/2 (2003): 377.

7 Shawn Achor, *The Happiness Advantage: The Seven Principles of Positive Psychology That Fuel Success and Performance at Work* (New York: Random House, 2011).

8 Anders Ericsson and Robert Pool, *Peak: Secrets from the New Science of Expertise* (New York: Houghton Mifflin Harcourt, 2016).

9 C. Hertzog and D. R. Touron, "Age Differences in Memory Retrieval Shift: Governed by Feeling-of-Knowing?" *Psychology and Aging* 26/3 (2011): 647–60.

10 D. R. Touron and C. Hertzog, "Age Differences in Strategic Behavior During a Computation-Based Skill Acquisition Task," *Psychology and Aging* 24/3 (2009): 574.

11 F. Sofi et al., "Physical Activity and Risk of Cognitive Decline: A Meta-Analysis of Prospective Studies," *Journal of Internal Medicine* 269/1 (2011): 107–17.

12 D. C. Park et al., "The Impact of Sustained Engagement on Cognitive Function in Older Adults: The Synapse Project," *Psychological Science* 25/1 (2013): 103–12.

13 Martin Samuels, "In Defense of Mistakes," *The Health Care Blog*, October 7, 2015, http://thehealthcareblog.com/blog/2015/10/07/in-defense-of-mistakes/.

14 Erica Klarreich, "How to Cut Cake Fairly and Finally Eat It Too," *Quanta Magazine*, October 6, 2016, https://www.quantamagazine.org/new-algorithm-solves-cake-cutting-problem-20161006/#.

15 Adam Grant, *Originals: How Non-Conformists Move the World* (New York: Penguin, 2016).

16 J. Boaler, K. Dance, and E. Woodbury, "From Performance to Learning: Assessing to Encourage Growth Mindsets," *youcubed*, 2018, https://bhi61nm2cr3mkdgk1dtaov18-wpengine.netdna-ssl.com/wp-content/uploads/2018/04/Assessent-paper-final-4.23.18.pdf.

17 Achor, *The Happiness Advantage*, 62–63.

18 Achor, *The Happiness Advantage*, 62–63.

Credits and Permissions

P. 64: "#TheLearningPit," from James Nottingham, *The Learning Challenge: How to Guide Your Students Through the Learning Pit to Achieve Deeper Understanding* (Thousand Oaks, CA: Corwin, 2017).

P. 83: Graph: growth mindset vs. fixed mindset, redrawn from L. S. Blackwell, K. H. Trzesniewski, and C. S. Dweck, "Implicit Theories of Intelligence Predict Achievement Across an Adolescent Transition: A Longitudinal Study and an Intervention," *Child Development* 78/1 (2007): 246–63.

P. 87: Graph: students receiving mindset workshop vs. students not receiving workshop, redrawn from L. S. Blackwell, K. H. Trzesniewski, and C. S. Dweck, "Implicit Theories of Intelligence Predict Achievement Across an Adolescent Transition: A Longitudinal Study and an Intervention," *Child Development* 78/1 (2007): 246–63.

P. 103: "Brain Networks for Mental Arithmetic," from V. Menon, "Salience Network," in Arthur W. Toga, ed., *Brain Mapping: An Encyclopedic Reference*, vol. 2 (London: Academic, 2015), 597–611.

P. 148: Concepts and methods schematic, redrawn from E. M. Gray and D. O. Tall, "Duality, Ambiguity, and Flexibility: A 'Proceptual' View of Simple Arithmetic," *Journal for Research in Mathematics Education* 25/2 (1994): 116–40.

INDEX